Chandramowli Subramanian

High Performance Computing for Stability Problems

Chandramowli Subramanian

High Performance Computing for Stability Problems

Applications to Hydrodynamic Stability and Neutron Transport Criticality

Südwestdeutscher Verlag für Hochschulschriften

Imprint
Any brand names and product names mentioned in this book are subject to trademark, brand or patent protection and are trademarks or registered trademarks of their respective holders. The use of brand names, product names, common names, trade names, product descriptions etc. even without a particular marking in this work is in no way to be construed to mean that such names may be regarded as unrestricted in respect of trademark and brand protection legislation and could thus be used by anyone.

Publisher:
Südwestdeutscher Verlag für Hochschulschriften
is a trademark of
Dodo Books Indian Ocean Ltd., member of the OmniScriptum S.R.L Publishing group
str. A.Russo 15, of. 61, Chisinau-2068, Republic of Moldova Europe
Printed at: see last page
ISBN: 978-3-8381-2689-0

Zugl. / Approved by: Karlsruhe Institute of Technology, Diss., 2011

Copyright © Chandramowli Subramanian
Copyright © 2011 Dodo Books Indian Ocean Ltd., member of the OmniScriptum S.R.L Publishing group

Contents

1. **Introduction** 1
   1.1 Function Spaces and Notations 3
2. **Hydrodynamic Stability** 5
   2.1 Modeling Fluid Flow Processes 6
      2.1.1 Conservation of Mass 7
      2.1.2 Balance of Momentum 8
      2.1.3 Constitutive Equations 9
      2.1.4 The Incompressible Navier-Stokes Equations 10
      2.1.5 Heat Conduction 10
      2.1.6 The Oberbeck-Boussinesq Equations 11
   2.2 Linear Stability Theory 12
      2.2.1 Stability of the Zero Solution 12
      2.2.2 Linear Stability of a Steady Flow 15
      2.2.3 The Bénard Problem 16
      2.2.4 Bibliographical Remarks 18
   2.3 Pseudospectra 18
      2.3.1 Pseudospectra of Matrices 18
      2.3.2 Pseudospectra of Linear Operators 21
      2.3.3 Bounds on Matrix and Operator Exponentials 22
      2.3.4 Pseudospectra of Matrix Pencils 27
3. **Neutron Transport Criticality** 29
   3.1 Neutron Transport 29
      3.1.1 General Assumptions 30
      3.1.2 Cross Section Definitions 31
      3.1.3 The Linear Boltzmann Equation 34
   3.2 Criticality 36
      3.2.1 The α Eigenvalue 36

		3.2.2 The k Eigenvalue	37

4 Discretization of the Eigenvalue Problems — 39
- 4.1 Galerkin Finite Element Spectral Approximation — 39
  - 4.1.1 Problem Formulation for Elliptic Eigenvalue Problems — 39
  - 4.1.2 Spectral Approximation of Compact Operators — 42
  - 4.1.3 A Priori Error Estimates for the Finite Element Approximation — 48
  - 4.1.4 Bibliographical Remarks — 54
- 4.2 Discretization of the Neutron Transport Equation — 54
  - 4.2.1 Energy Discretization — 54
  - 4.2.2 Second Order Even Parity Formulation — 56
  - 4.2.3 Angular and Spatial Discretization — 58

5 Eigenvalue Solvers and Parallelization — 61
- 5.1 The Davidson Method — 61
- 5.2 Computation of Pseudospectra — 65
- 5.3 Parallelization — 67
  - 5.3.1 Sparse Matrix Vector Multiplication for Finite Element Methods — 67
  - 5.3.2 Parallel Computation of Pseudospectra — 69

6 Pseudospectra in Hydrodynamic Stability — 71
- 6.1 Incompressible Flow over a Backward Facing Step — 71
- 6.2 Natural Convection in a Horizontal Annulus — 80

7 Applications in Nuclear Reactor Theory — 85
- 7.1 The Takeda 1 Benchmark — 86
- 7.2 The NEA C5G7 Benchmark — 91

8 Summary and Outlook — 95

A Estimates and Calculations for the Poincaré Constant — 97
- A.1 Bounds for the Poincaré Constant — 97
- A.2 Evaluation of the Poincaré Constant — 101
- A.3 Application in Natural Convection — 101

Bibliography — 105

Chapter 1

Introduction

The question of stability is crucial in a variety of disciplines, such as engineering, control theory, physics, and mathematics. This can be seen in a large number of applications: designing bridges or buildings, controlling nuclear reactors, stabilizing flow processes in pipelines, developing electronic stability control for vehicles, or avoiding turbulences around aircrafts. To cope with these kinds of problems is an enormous challenge due to the high complexity of the whole solution process. First, an adequate mathematical model describing the underlying physical processes needs to be constituted. Then, appropriate discretizing techniques able to approximate the model accurately have to be established. In order to solve the resulting large-scaled problems in a reasonable time, one further needs to employ efficient numerical methods capable of exploiting parallel platforms.

In this work we treat two different stability problems. First, we devote ourselves to the stability of fluid flow processes, which is a major field in the theory of fluid dynamics, namely the *hydrodynamic stability theory*. In this respect, the examination is based on exerting a perturbation on a laminar flow, and then to study the evolution of the perturbed quantity. By means of eigenvalue problems, we determine whether the perturbation decays or grows in time. We focus on the linear stability analysis which is directly linked to the spectrum of the linearized operator describing the flow. Unlike a nonlinear stability analysis, a purely linear stability analysis cannot guarantee the stability of the underlying process. This work discusses the applicability of *pseudospectra* to tackle this matter.

Pseudospectra are also based on the linearized problem. Despite of the quite expensive evaluation of pseudospectra compared to spectra, we believe that such an investigation is valuable, as in the advent of high performance computing more computational power is available, and by means of pseudospectra, one may gain a deeper insight in the stability behavior of physical systems.

Based on results already available to approximate eigenvalues of elliptic operators, we derive the mathematical foundation to approximate pseudospectra as well. Our general approach allows us to treat complex flow processes approximated by finite element methods. As for the computation, which consists of evaluating singular values, we focus on the *Davidson method* and propose an efficient computational scheme for parallel hardware architectures.

The second aspect treated in this work is the *criticality* problem arising in the field of neutron transport theory. It is concerned with the stability of nuclear fission chain-reactions. The typical application is the modeling and controlling of nuclear reactors. The criticality problem may also be formulated by means of an eigenvalue problem. The largest eigenvalue of the resulting problem indicates how far the fission chain-reaction is from the desired state of equilibrium, where the amount of neutrons emitted equals the amount of neutrons absorbed.

So far, the *power method* has been the method of choice to solve the criticality eigenvalue problem. Some more elaborated methods, such as the *Arnoldi method* or the *Jacobi-Davidson method*, have also been successfully applied to this problem. However, to the best of the authors' knowledge, these methods have only been used in the context of some simpler approximation schemes (P_1 and S_N methods to resolve the angular dependence, see Section 4.2).

In this work we focus on the *k eigenvalue problem* which is one of the widely used formulation to state the criticality problem. We apply the Davidson method to solve this generalized eigenvalue problem in the framework of a more sophisticated approximation scheme (P_N method). Our numerical results show that this method is a promising alternative to the rather simple power method by speeding up our calculations significantly. Furthermore, the Davidson method shows more flexibility and robustness than the power method.

This book is organized as follows. In the following Section 1.1 we establish some basic definitions and notations needed throughout this work. In Chapter 2 we explain the concepts of hydrodynamic stability. We start by establishing the equations describing incompressible flow and natural convection processes. Afterwards, we focus on linear stability theory and the utilization of pseudospectra in this respect. The subject of Chapter 3 is to describe the linear Boltzmann equation modeling the neutron transport, and to formulate eigenvalue problems describing the criticality. Chapter 4 is concerned with the theoretical background of the discretization methods which are used to approximate the models arising in the hydrodynamic stability theory and the criticality problem. In both applications we employ the Davidson method which is treated in Chapter 5. Moreover, this chapter illustrates the basic mechanism of the applied parallelization techniques. In Chapter 6 we present pseudospectra of two different flow processes: We consider an incompressible flow over a backward facing step as well as a natural convection process in an annulus. Numerical results for criticality problems are shown in Chapter 7. In Chapter 8 we discuss our results and give an outlook to potential future research fields. Additionally, in Appendix A we derive a bound for the *Poincaré constant* in an annulus and check its quality numerically. Moreover, we apply the obtained results to a natural convection problem.

1.1 Function Spaces and Notations

We give a short introduction to Lebesgue and Sobolev spaces, as we need these definitions throughout this work, especially for the finite element framework. These definitions are rather standard and can for instance be found in [21, 50, 93].

Suppose Ω to be a Lebesgue measurable subset of $\mathbb{R}^d$ (usually $d = 2$ or $d = 3$) with non-empty interior. Let f be a real or complex valued function on Ω which is Lebesgue measurable. By
$$\int_\Omega f(x)\,dx$$
we denote the Lebesgue integral of f over Ω. For $1 \leq p \leq \infty$ we define the *Lebesgue norm*:

$$\|f\|_{L^p(\Omega)} = \begin{cases} \left(\int_\Omega |f(x)|^p\,dx\right)^{1/p}, & 1 \leq p < \infty, \\ \operatorname*{ess\,sup}_{x \in \Omega} |f(x)| = \inf\{C \geq 0 : |f(x)| < C \text{ almost everywhere}\}, & p = \infty. \end{cases}$$

It can be shown that the *Lebesgue spaces* $L^p(\Omega) = \{f : \|f\|_{L^p(\Omega)} < \infty\}$ equipped with the norm $\|\cdot\|_{L^p(\Omega)}$ are Banach spaces for any $1 \leq p \leq \infty$. In this context, we identify two functions f and g if $f(x) = g(x)$ almost everywhere, i.e. $\|f - g\|_{L^p(\Omega)} = 0$. For the special case $p = 2$, we have that $L^2(\Omega)$ is a Hilbert space with the inner product

$$(f,g)_0 := (f,g)_{L^2(\Omega)} = \int_\Omega f(x)\overline{g(x)}\,dx. \tag{1.1}$$

Let $L^1_{loc} = \{f \in L^1(K) : K \text{ compact}, K \subset \mathring{\Omega}\}$ denote the set of *locally integrable* functions. Here, $\mathring{\Omega}$ refers to the interior of Ω. Furthermore, $C_0^\infty(\Omega)$ is the set of functions in $C^\infty(\Omega)$ with compact support in Ω. Let D^α be shorthand for the partial derivative operator:
$$D^\alpha := \frac{\partial^{|\alpha|}}{\partial x_1^{\alpha_1}\,\partial x_2^{\alpha_2}\,\cdots\,\partial x_n^{\alpha_d}}.$$
Here, the n-tuple $\alpha = (\alpha_1, \alpha_2, \ldots, \alpha_n)$ is a *multi-index* with $\alpha_i \in \mathbb{N}_0$ and $|\alpha| = \sum_{i=1}^d \alpha_i$.

We define a *weak derivative* $D_w^\alpha f$ of a function $f \in L^1_{loc}(\Omega)$ if we have that $D_w^\alpha f \in$

$L^1_{loc}(\Omega)$ and
$$\int_\Omega D_w^\alpha f(x)\varphi(x)\,dx = (-1)^{|\alpha|}\int_\Omega f(x)D^\alpha\varphi(x)\,dx$$
holds for any $\varphi \in C_0^\infty(\Omega)$. Since for any function $\varphi \in C^{|\alpha|}(\Omega)$ the classical derivative coincides with the weak derivative, we do not make any distinction in the notation of D and D_w.

Equipped with the definition of weak derivatives, we can set up the framework for Sobolev spaces. For any $k \in \mathbb{N}_0$ we set the *Sobolev norm*
$$\|f\|_{W_p^k(\Omega)} = \begin{cases} \left(\sum_{|\alpha|\leq k}\|D^\alpha f\|_{L^p(\Omega)}^p\right)^{1/p}, & 1 \leq p < \infty, \\ \max_{|\alpha|\leq k}\|D^\alpha f\|_{L^\infty(\Omega)}^p, & p = \infty. \end{cases}$$

The *Sobolev spaces* defined by $W_p^k(\Omega) = \{f \in L^1_{loc}(\Omega) : \|f\|_{W_p^k(\Omega)} < \infty\}$ endowed with the corresponding Sobolev norm are Banach spaces. For the case $p = 2$, $H^k(\Omega) := W_2^k(\Omega)$ are Hilbert spaces with the inner product
$$(f,g)_k := (f,g)_{H^k(\Omega)} = \sum_{|\alpha|\leq k}(D^\alpha f, D^\alpha g)_{L^2(\Omega)}.$$

Note that for $k = 0$, we have $H^0(\Omega) = L^2(\Omega)$ with the same inner product $(\cdot,\cdot)_0$ as in (1.1). Moreover, we define $H_0^k(\Omega)$ to be the completion of $C_0^\infty(\Omega)$ with respect to the norm $\|\cdot\|_{H^k(\Omega)}$.

For a vector $\mathbf{x}$ we set $\mathbf{x}^H = \overline{\mathbf{x}^T}$ (the complex conjugate of the transposed vector). The same notation is used for a matrix $\mathbf{A}$, i.e. $\mathbf{A}^H = \overline{\mathbf{A}^T}$. Finally, for two vectors $\mathbf{x}, \mathbf{y}$ in $\mathbb{R}^d$ or $\mathbb{C}^d$, we denote the usual scalar product by $\mathbf{x}\cdot\mathbf{y} = \mathbf{x}^H\mathbf{y}$.

Chapter 2

Hydrodynamic Stability

Stability of a fluid flow is a crucial question in many applications affecting our daily life: How stable is a flying aircraft exposed to gusting winds? Can particles in pipelines, transporting oil or gas, cause congestions? How can turbulences in a liquid, transporting the heat out of a nuclear reactor, be avoided in order to maintain a failure-free cooling system?

These kinds of questions are very complex. For a theoretical analysis, an appropriate model describing the underlying phenomena needs to be constituted. In most cases, many assumptions have to be made in order to get a formulation (e.g. partial differential equations) both pure and numerical analysis can cope with. But no matter which model we choose, just few aspects of the complex flow behavior which is observed are reflected by available theoretical results – mostly for very simple configurations.

The approach of hydrodynamic stability is to investigate how a laminar fluid flow behaves with respect to perturbations. If the perturbation decays in time and the flow returns to its original state it is said to be stable. On the other hand, if the perturbation causes the flow to change into a different state, it is said to be instable. Instability may trigger turbulence, but it may also take the flow to a different laminar state. Instability means in particular, a computed solution might not be observable in experiments or the control of a procedure in industrial production is just impossible.

In the field of hydrodynamic stability there are two main approaches. The *nonlinear stability* theory is based on examining the kinetic energy of the flow by means of integral inequality techniques. This method is also referred to as *energy method*. The *linear stability* theory is concerned with a linearized model of the fluid flow and establishes statements by means of the spectrum of the linearized operator. If all eigenvalues lie in the left half of the complex plane, the flow is said to be linear stable. If there is at least one eigenvalue in the right half of the complex plane, the flow is linear instable (and therefore instable as we will see).

In Section 2.1 we give a brief review of the governing equations modeling fluid flow processes, where we consider two different types. First, we study an incompressible fluid flow modeled by the Naiver-Stokes equations. Afterwards, we extend this model in order to consider a heat driven flow. This phenomenon is formulated by means of the Oberbeck-Boussinesq equations. Subject of the succeeding Section 2.2 is to establish the basic definitions of linear stability theory in terms of the Navier-Stokes equations. Moreover, we point out the differences between linear and nonlinear stability and show the close relation between stability and eigenvalue problems. Afterwards, in Section 2.3 we give an outline to a more general concept than eigenvalues, namely pseudospectra. We describe how pseudospectra can be an asset to gain deeper insights in the stability behavior of physical systems.

2.1 Modeling Fluid Flow Processes

We start in this section by deriving the basic equations which describe flow processes of incompressible liquids or gases, namely the *incompressible Navier-Stokes equations*, see e.g. [7, 28, 82, 114]. Afterwards, we are also concerned with flow processes governed by temperature differences and gravity forces, the so-called *Oberbeck-Boussinesq equations*. Both models are constituted by the conservation laws of mass and momentum, whereas the Oberbeck-Boussinesq equations additionally require the balance of energy to formulate the heat conduction.

Let $\Omega \in \mathbb{R}^d$ ($d = 2$ or $d = 3$) be a region filled with a fluid moving due to prevailing internal and external forces. In what follows, we consider a bounded *test volume* $V(t) \in \Omega$ which satisfies the requirements to apply the divergence theorem. Furthermore, we denote d-dimensional vectors by bold-faced letters, e.g. $\mathbf{x} \in \Omega$ denotes the spatial variable in $\mathbb{R}^d$. For each time t, we assume that the fluid has a well-defined and continuous *mass density* $\rho(\mathbf{x}, t) > 0$ such that the mass $m(t)$ of the volume $V(t)$ is given by

$$m(V(t)) = \int_{V(t)} \rho(\mathbf{x}, t) \, d\mathbf{x}. \qquad (2.1)$$

This condition is also referred to as *continuum hypothesis*. The continuum hypothesis allows us to speak of physical properties of an infinitesimally small volume element of the fluid, which we call *material particle* in what follows. We define the movement of a material particle $\eta \in V(t)$ by a function $\mathbf{x}(\eta, t)$ such that the material particle η at time t is located at position $\mathbf{x}(\eta, t) \in \Omega$. We assume the function $\mathbf{x}$ to be invertible and furthermore to be sufficiently smooth.

A flow may be described by the physical properties of each material particle as a

2.1. Modeling Fluid Flow Processes

function of time, which is called *Lagrangean formulation*, i.e. for any fixed material particle η one follows its trajectory. Whereas in the *Eulerian formulation* the flow is described by specifying the physical properties as a function of time for any fixed point of the domain. In the framework of the Eulerian formulation, we define the velocity of a material particle which is at position $\mathbf{x}(\eta, t)$ by the d-dimensional vector field

$$\mathbf{v}(\mathbf{x}, t) = \partial_t \mathbf{x}(\eta, t).$$

Let $f(\mathbf{x}, t)$ represent any scalar physical property of the fluid at position $\mathbf{x}$ and time t. An observation of the temporal change of f at a fixed point in a chosen coordinate system is described by the *local derivative* $\partial_t f(\mathbf{x}, t)$. Observing the temporal variation of f for a fixed particle η by $f(\mathbf{x}, t) = f(\mathbf{x}(\eta, t), t)$ leads to the time derivative in the Lagrangean formulation, the so-called *material derivative*:

$$\frac{df}{dt} = \frac{df}{dt}(\mathbf{x}(\eta, t), t) = \mathbf{v} \cdot \nabla f + \partial_t f,$$

where the nabla operator ∇ acts only on the spatial variable $\mathbf{x}$. Equipped with these definitions, we are able to state the *Transport theorem*. Therefore, we assume that f, and other functions to be introduced later, are smooth enough to apply standard operations on them.

Theorem 2.1 (Transport theorem) *For any smooth scalar field $f(\mathbf{x}, t)$ and any test volume $V(t)$ we have*

$$\frac{d}{dt} \int_{V(t)} f(\mathbf{x}, t) \, d\mathbf{x} = \int_{V(t)} \partial_t f(\mathbf{x}, t) + \nabla \cdot (f(\mathbf{x}, t) \mathbf{v}(\mathbf{x}, t)) \, d\mathbf{x}.$$

For a proof, see e.g. [28, 114].

The derivation of the Navier-Stokes equations is based on the following conservation laws, which are discussed in the following sections.

(1) Conservation of mass: Mass is neither created nor destroyed.

(2) Conservation of momentum (*Newton's second law*): The momentum change rate of a material volume is equal to the force exerted.

2.1.1 Conservation of Mass

The mass conservation law states that the mass $m(V(t))$ of an arbitrary test volume is constant in time, i.e.

$$\frac{d}{dt} m(V(t)) = \frac{d}{dt} \int_{V(t)} \rho(\mathbf{x}, t) \, d\mathbf{x} = 0.$$

Applying the Transport theorem 2.1, we deduce that

$$\int_{V(t)} \partial_t \rho(\mathbf{x},t) + \nabla \cdot (\rho(\mathbf{x},t)\mathbf{v}(\mathbf{x},t)) \, d\mathbf{x} = 0$$

holds for any $V(t)$. Assuming the integrand to be continuous, we have that at each point

$$\partial_t \rho + \nabla \cdot (\rho \mathbf{v}) = 0, \tag{2.2}$$

which is known as the *continuity equation*.

2.1.2 Balance of Momentum

For any continuum, there are two types of forces acting on a piece of material $V(t)$. *Body* (or *external*) *forces*, such as gravity or electromagnetic forces, can be regarded as acting throughout a volume. These are quantified by

$$F_{ext}(V(t)) = \int_{V(t)} \rho(\mathbf{x},t) \mathbf{f}(\mathbf{x},t) \, d\mathbf{x},$$

with the vector valued function $\mathbf{f}(\mathbf{x},t)$ representing an external force density. Whereas *contact* (or *internal*) *forces* act through the surface of $V(t)$ and are described by the *stress vector* $\mathbf{s}(\mathbf{x},t)$ acting per unit area:

$$F_{int}(V(t)) = \int_{\partial V(t)} \mathbf{s}(\mathbf{x},t) \, ds.$$

Here, $\partial V(t)$ refers to the boundary of the test volume $V(t)$. Let $\mathbf{n}$ denote the outward unit normal. It can be shown that $\mathbf{s}(\mathbf{x},t) = \mathbf{n} \cdot \sigma$, where $\sigma = \sigma_{ij}$ is the *stress tensor*, see e.g. [7, 82].

The linear momentum conservation law states that

$$\frac{d}{dt} \int_{V(t)} \rho(\mathbf{x},t) \mathbf{v}(\mathbf{x},t) \, d\mathbf{x} = \int_{V(t)} \rho(\mathbf{x},t) \mathbf{f}(\mathbf{x},t) \, d\mathbf{x} + \int_{\partial V(t)} \mathbf{s}(\mathbf{x},t) \, ds \tag{2.3}$$

holds for any test volume $V(t)$. Applying the transport theorem and the divergence theorem, and, moreover, using the continuity equation (2.2) yields

$$\rho \partial_t \mathbf{v} + \rho (\mathbf{v} \cdot \nabla) \mathbf{v} = \rho \mathbf{f} + \nabla \cdot \sigma. \tag{2.4}$$

This resulting equation (2.4) is called *non-conservative form of the momentum equation*.

Using the principle of conservation of angular momentum, one can show that the stress tensor is symmetric, i.e. $\sigma^T = \sigma$, see e.g. [7, 82].

2.1.3 Constitutive Equations

So far, the derived equations are generic and appropriate for all kinds of fluids. However, the number of unknowns

- velocity vector $\mathbf{v} \in \mathbb{R}^d$,
- density $\rho \in \mathbb{R}$,
- stress tensor $\sigma \in \mathbb{R}^{d \times d}$

does not match the number of conditions imposed by the following equations

- continuity equation: $\partial_t \rho - \nabla \cdot (\rho \mathbf{v}) = 0$,
- momentum equation $\rho \partial_t \mathbf{v} + \rho (\mathbf{v} \cdot \nabla) \mathbf{v} = \rho \mathbf{f} + \nabla \cdot \sigma$,
- conservation of angular momentum: $\sigma = \sigma^T$.

In order to complete the system, we need to relate the stress tensor to the velocity and density. These relations are derived empirically in physical experiments. The first assumption we make, is that the fluid is *Stokesian*, i.e. the stress tensor is spherically symmetric when the fluid is at rest:

$$\sigma|_{\mathbf{v}=0} = -pI,$$

where $p = p(\mathbf{x}, t)$ denotes the *thermodynamic pressure*. This means, that the shear stress components of σ vanish, while the normal stresses are equal to the pressure. Apparently, for a fluid in motion this is not the case. Here, we set

$$\sigma|_{\mathbf{v} \neq 0} = -pI + \tau, \qquad (2.5)$$

where τ is a symmetric tensor describing the shear stresses. To get a relation for τ we define the *deformation* (or *rate of strain*) *tensor* as $D = \frac{1}{2} \left(\nabla \mathbf{v} + (\nabla \mathbf{v})^T \right)$. We assume that $\tau = F(D)$, where F is an appropriate continuous function mapping symmetric tensors on symmetric tensors. If we furthermore assume F to be linear (i.e. we consider *Newtonian fluids*) and isotropic, equation (2.5) yields

$$\sigma = (-p + \lambda \nabla \cdot \mathbf{v}) I + 2\mu D,$$

see [7, 28, 82]. Here, λ is the *volume viscosity* and μ the *shear viscosity*.

2.1.4 The Incompressible Navier-Stokes Equations

Incompressibility means that the density is constant in space and time. Thus, by virtue of the continuity equation (2.2) the velocity is solenoidal, i.e. $\nabla \cdot \mathbf{v} = 0$. This implies

$$\begin{aligned}\nabla \cdot \sigma &= -\nabla p + \nabla\left(\lambda \nabla \cdot \mathbf{v}\right) + 2\mu \nabla \cdot D \\ &= -\nabla p + \mu\left(\Delta \mathbf{v} + \nabla(\nabla \cdot \mathbf{v})\right) \\ &= -\nabla p + \mu \Delta \mathbf{v}.\end{aligned}$$

Summing up, with (2.4) and (2.2), and defining the *kinematic viscosity* as $\nu = \mu/\rho$, we obtain the *incompressible Navier-Stokes equations*:

$$\begin{aligned}\partial_t \mathbf{v} - \nu \Delta \mathbf{v} + (\mathbf{v} \cdot \nabla)\mathbf{v} + \frac{1}{\rho}\nabla p &= \mathbf{f}, \\ \nabla \cdot \mathbf{v} &= 0,\end{aligned} \quad (2.6)$$

where suitable initial and boundary conditions on $\mathbf{v}$ have to be imposed.

2.1.5 Heat Conduction

Up to now, we have studied only *isothermal* flow processes, i.e. we assumed a constant temperature, whereas in this section we introduce flow processes with heat variation. Typical applications include cooling systems and insulation technologies.

Thermal convections, where the convection is only driven by some external forces (e.g. by a fan), and the effects of buoyancy are neglected, are known as *forced convections*. However, no new flow phenomena arise, since the motion of the fluid is unaffected by the temperature and can be determined in the way as in the preceding Section 2.1.4. On the other hand, the case where the convection is only governed by buoyancy forces is called *natural convection*. In this case the fluid would be at rest if no temperature variation was introduced. In many applications the motion of the fluid is induced by temperature difference between boundaries. *Mixed convection* is the combination of forced and natural convection, i.e. buoyancy forces and external forces are both considered.

In this work, we confine ourselves to natural convection, where the fluid is moving due to temperature differences and gravity. This phenomenon is formulated by means of the *Oberbeck-Boussinesq equations* (or *Boussinesq approximation*). These equations are based on the continuity equation, the momentum equation, and an additional equation for the heat conduction.

So far, we have deduced the equations of continuity and motion by using the laws of conservation of mass and momentum respectively. To formulate the heat conduction in a fluid, we need to exploit the conservation of energy as well which is expressed by the first

2.1. Modeling Fluid Flow Processes

law of thermodynamics. The equation of heat conduction reads

$$\rho \, \partial_t(c_V \, T) + \rho \, \mathbf{v} \cdot \nabla(c_V \, T) = \nabla \cdot (k \, \nabla T) - p \, \nabla \cdot \mathbf{v} + \Phi, \qquad (2.7)$$

see [26, 105, 114]. Here, $T = T(\mathbf{x}, t)$ denotes the temperature. Furthermore, c_V is the *specific heat* at constant volume, k is the *thermal conductivity*. The *energy dissipation* Φ per unit volume by viscosity reads in the case of Newtonian fluids

$$\Phi = 2\mu D : D - \frac{2}{3}(\nabla \cdot \mathbf{v})^2,$$

where the scalar product of two tensors is defined by $A : B = \sum_{ij} A_{ij} B_{ij}$.

2.1.6 The Oberbeck-Boussinesq Equations

Boussinesq [18] and Oberbeck [73] recognized that if the temperature variations are small, the dynamics of a fluid can be approximated by assuming a constant density everywhere, except in the buoyancy. To give a brief sketch of this argumentation, suppose that the density can be quantified by the linear dependence

$$\rho = \rho_0 \left[1 + \alpha(T_0 - T)\right],$$

where T_0 denotes the temperature for which $\rho = \rho_0$ and α is the *volumetric expansion coefficient*. Usually, α is in the range of 10^{-3} to 10^{-4} K^{-1} (e.g. perfect gas: $\alpha \approx 3 \cdot 10^{-3}$ K^{-1}; liquids which are mostly used in experiments: $\alpha \approx 5 \cdot 10^{-4}$ K^{-1}). If we have small temperature differences, say up to 10 K, the variations in the density are at most one per cent. That is why the density variations may be ignored in all terms of (2.2), (2.4) and (2.7), with one exception: the term describing the gravity force, see [26, 38, 105]. This external force is given by $f_g = \rho \mathbf{g}$, where $\mathbf{g}$ is the *acceleration due to gravity*.

Next, we have to simplify equation (2.7). Since we have replaced the continuity equation by $\nabla \cdot \mathbf{v} = 0$, we can ignore the term $p \nabla \cdot \mathbf{v}$. Furthermore, we can treat c_V and k as constants. Finally, we neglect the viscous dissipation Φ since it is most commonly of very low order. Summing up, the Oberbeck-Boussinesq equations read

$$\partial_t \mathbf{v} - \nu \Delta \mathbf{v} + (\mathbf{v} \cdot \nabla) \mathbf{v} + \frac{1}{\rho} \nabla p = \mathbf{g} \left[1 + \alpha(T_0 - T)\right],$$
$$\nabla \cdot \mathbf{v} = 0, \qquad (2.8)$$
$$\partial_t T + \mathbf{v} \cdot \nabla T - \kappa \Delta T = 0,$$

where we have replaced ρ_0 by ρ and $\kappa = k/(\rho c_V)$ denotes the *thermal diffusity*. For both, the velocity $\mathbf{v}$ and the temperature distribution T, suitable initial and boundary conditions have to be prescribed. A more detailed physical background describing the approximations we used in this section can be found in [105].

2.2 Linear Stability Theory

In this section we show the close relation between stability analysis and eigenvalue problems for the incompressible Navier-Stokes equations. Furthermore, we briefly discuss the stability of the Bénard problem. Later in this work, in Section 2.3.3, the linear stability theory is put in a more abstract framework which allows us to establish bounds by means of spectra and pseudospectra.

For simplicity, we scale the pressure p by setting $\rho = 1$. Then, the viscous fluid flow we consider is governed by

$$\partial_t \mathbf{v} - \nu \Delta \mathbf{v} + (\mathbf{v} \cdot \nabla) \mathbf{v} + \nabla p = \mathbf{f},$$
$$\nabla \cdot \mathbf{v} = 0 \tag{2.9}$$

in a bounded domain Ω, with boundary conditions $\mathbf{v} = \mathbf{h}$ on $\partial\Omega$ and an initial value $\mathbf{v}(\mathbf{x}, 0) = \mathbf{v}_0(\mathbf{x})$.

Before we go deeper into the linear stability theory, we start with an example to show the different approaches of linear and the nonlinear stability. In what follows, we assume $\mathbf{v}$ and p to be sufficiently smooth and the domain Ω under consideration to fulfill the requirements of the divergence theorem.

2.2.1 Stability of the Zero Solution

In this example we examine the stability of the zero solution of (2.9), where the external force $\mathbf{f}$ is given by a potential $\mathbf{f} = -\nabla \Phi$. Moreover, we impose an initial value $\mathbf{v}(\mathbf{x}, 0) = \mathbf{v}_0(\mathbf{x})$ for $\mathbf{x} \in \Omega$, and zero boundary values, i.e. $\mathbf{v}(\mathbf{x}, t) = \mathbf{0}$ for any $\mathbf{x} \in \partial\Omega$. Physically, this setup corresponds to a model of a container filled with a liquid which is arbitrarily moved around for $t < 0$, and then held fixed at $t = 0$. For $t \geq 0$ the container is only exposed to the gravitational force $-\nabla \Phi$. Empirically, we expect the perturbation to decay and the liquid to return to its initial state.

Nonlinear Stability

The nonlinear stability considers the *kinetic energy* $E(t)$ which is given by

$$E(t) = \frac{1}{2}\|\mathbf{v}\|_0^2 = \frac{1}{2}(\mathbf{v}, \mathbf{v})_0.$$

2.2. Linear Stability Theory

Here, $(\cdot, \cdot)_0$ is the inner product in $L^2(\Omega)$ and $\|\cdot\|_C$ the corresponding $L^2(\Omega)$ norm, cf. Section 1.1. Differentiating $E(t)$ with respect to t yields

$$\frac{d}{dt}E(t) = (\partial_t \mathbf{v}, \mathbf{v})_0.$$

Next, the first equation of (2.9) is multiplied by $\mathbf{v}$ and we take the integral over Ω. Then we integrate by parts to obtain

$$\frac{d}{dt}E(t) = -\nu \|\nabla \mathbf{v}\|_0^2 + \frac{1}{2}(\mathbf{v} \cdot \mathbf{v}, \nabla \cdot \mathbf{v})_0 + (p + \Phi, \nabla \cdot \mathbf{v})_0,$$

see [94]. By assumption $\mathbf{v}$ is divergence free which implies

$$\frac{d}{dt}E(t) = -\nu \|\nabla \mathbf{v}\|_0^2.$$

From the Poincaré inequality (A.1) we deduce

$$\frac{d}{dt}E(t) \leq -\nu k_p^2 \|\mathbf{v}\|_0^2 = -2\nu k_p^2 \, E(t),$$

hence

$$E(t) \leq e^{-2\nu k_p^2 t} E(0).$$

Since $k_p > 0$, we have that $\|\mathbf{v}\|_0 \to 0$ as $t \to \infty$, which means that all initial perturbations die away. Note that the kinematic viscosity ν as well as the Poincaré constant k_p determine the evolution of the disturbance.

Linear Stability

The linear stability theory follows a different approach. We assume the perturbation to be small in order to neglect products of higher order containing $\mathbf{v}$. This means, we consider the linear problem

$$\partial_t \mathbf{v} - \nu \Delta \mathbf{v} + \nabla p = \mathbf{f},$$
$$\nabla \cdot \mathbf{v} = 0$$

in Ω, with the same initial and boundary conditions as before. Suppose that $\mathbf{v}$ and p are comprised of superpositions of *normal modes*

$$\mathbf{v}(\mathbf{x}, t) = \tilde{\mathbf{v}}(\mathbf{x}) e^{\lambda t}, \quad p(\mathbf{x}, t) = \tilde{p}(\mathbf{x}) e^{\lambda t},$$

where the eigenvalues λ and the corresponding eigenfunctions $(\tilde{\mathbf{v}}, \tilde{p})$ satisfy

$$\lambda \tilde{\mathbf{v}} = \nu \Delta \tilde{\mathbf{v}} - \nabla \tilde{p}, \\ 0 = \nabla \cdot \tilde{\mathbf{v}} \quad (2.10)$$

in Ω, with zero Dirichlet boundary conditions for $\tilde{\mathbf{v}}$. Clearly, any mode with $\mathrm{Re}\,\lambda < 0$ decays in time, whereas a mode with $\mathrm{Re}\,\lambda > 0$ grows exponentially. Suppose we have countably many eigenvalues with no accumulation point at 0. This can be shown in the framework of elliptic operators assuming a bounded domain Ω with a sufficiently smooth boundary $\partial \Omega$, see Section 4.1. We say that the zero solution is *linear stable* if all normal modes decay, i.e. $\mathrm{Re}\,\lambda < 0$ for all eigenvalues. On the other hand, if there exists one eigenvalue λ with $\mathrm{Re}\,\lambda > 0$, the zero solution is said to be *linear instable*, see [37, 38].

In our case, it is easily shown that $\mathrm{Re}\,\lambda < 0$ holds for all modes. We multiply the first equation of (2.10) by $\tilde{\mathbf{v}}$ and integrate over Ω. Then integrating by parts under consideration of $\nabla \cdot \tilde{\mathbf{v}} = 0$ leads to

$$\lambda \|\tilde{\mathbf{v}}\|_0^2 = -\nu \|\nabla \tilde{\mathbf{v}}\|_0^2.$$

As a consequence of the Poincaré inequality (A.1) we obtain

$$\lambda \leq -\nu k_p^2 < 0$$

for all eigenvalues λ. Note that, as in the case of nonlinear stability, the Poincaré constant and the kinematic viscosity are essential for the stability. Both quantities prescribe the evolution of the normal modes.

As we have seen exemplarily for the stability of the zero solution, both approaches, linear and nonlinear stability, are closely related to eigenvalue problems. The energy method employs the Poincaré constant, which can be determined by means of an eigenvalue problem (for bounded domains), see Appendix A, whereas the linear stability theory uses the spectrum of the operator in addition. The nonlinear stability theory can guarantee stability, since it provides an upper bound for the evolution of the perturbation. However, if the upper bound tends to infinity, no statement in terms of stability can be made. On the other hand, the linear stability theory can assure instability as soon as we have one eigenvalue lying in the right half of the complex plane. But linear stability does not imply nonlinear stability in general.

2.2.2 Linear Stability of a Steady Flow

To investigate the stability of an arbitrary steady flow $(\mathbf{V}(\mathbf{x}), P(\mathbf{x}))$ governed by (2.9), we disturb the steady solution at $t = 0$ by a function $(\mathbf{v}_0, p_0)$. The selected basic flow $(\mathbf{V}, P)$ is assumed to be known, either analytically or only by means of numerical computations. In order to study the evolution of the perturbation $(\mathbf{v}(\mathbf{x}, t), p(\mathbf{x}, t))$, we insert the perturbed quantities $(\mathbf{V} + \mathbf{v}, P + p)$ in (2.9) yielding

$$\partial_t \mathbf{v} - \nu \Delta \mathbf{v} + (\mathbf{v} \cdot \nabla)\mathbf{V} + (\mathbf{V} \cdot \nabla)\mathbf{v} + (\mathbf{v} \cdot \nabla)\mathbf{v} + \nabla p = 0,$$
$$\nabla \cdot \mathbf{v} = 0 \quad (2.11)$$

in Ω, with zero boundary conditions $\mathbf{v} = \mathbf{0}$ on $\partial \Omega$ and initial conditions $\mathbf{v}(\mathbf{x}, 0) = \mathbf{v}_0$, $p(\mathbf{x}, 0) = p_0$.

We follow the stability theory of ordinary differential equations by defining stability in the sense of Lyapunov, see [4, 37, 51]. Therefore, we assume the functions $\mathbf{v}$ and p under consideration to be elements of a Banach space, where, for brevity, we denote all norms by $\|\cdot\|$. A basic flow $(\mathbf{V}, P)$ is said to be *stable* if for all $\varepsilon > 0$ there exists a $\delta = \delta(\varepsilon)$ such that

$$\|\mathbf{v}(\mathbf{x}, 0)\| < \delta \quad \text{and} \quad \|p(\mathbf{x}, 0)\| < \delta$$

implies

$$\|\mathbf{v}(\mathbf{x}, t)\| < \varepsilon \quad \text{and} \quad \|p(\mathbf{x}, t)\| < \varepsilon \quad \text{for all } t > 0.$$

Otherwise it is said to be *instable*. This means, a stable flow ensures that all initially small perturbations remain small for all time. If, additionally, the perturbations decay asymptotically, we say that the basic flow is *asymptotically stable*, i.e. it is stable and

$$\|\mathbf{v}(\mathbf{x}, t)\| \to 0 \quad \text{and} \quad \|p(\mathbf{x}, t)\| \to 0 \quad \text{as } t \to \infty.$$

The linear stability theory assumes the perturbations to be small. So we may neglect products of the perturbed quantities in (2.11) yielding the linear problem

$$\partial_t \mathbf{v} - \nu \Delta \mathbf{v} + (\mathbf{v} \cdot \nabla)\mathbf{V} + (\mathbf{V} \cdot \nabla)\mathbf{v} + \nabla p = 0,$$
$$\nabla \cdot \mathbf{v} = 0, \quad (2.12)$$

with the same initial and boundary conditions. Since the basic flow is steady, the coefficients in (2.12) are independent of t. Suppose that a solution of the initial value problem (2.12) may be separated, i.e. it is a linear superposition of *normal modes*

$$\mathbf{v}(\mathbf{x}, t) = \tilde{\mathbf{v}}(\mathbf{x}) e^{\lambda t}, \quad p(\mathbf{x}, t) = \tilde{p}(\mathbf{x}) e^{\lambda t},$$

where the eigenvalues λ and corresponding eigenfunctions $(\tilde{\mathbf{v}}, \tilde{p})$ satisfy

$$\lambda \tilde{\mathbf{v}} = \nu \Delta \tilde{\mathbf{v}} - (\tilde{\mathbf{v}} \cdot \nabla)\mathbf{V} - (\mathbf{V} \cdot \nabla)\tilde{\mathbf{v}} - \nabla \tilde{p}, \qquad (2.13)$$
$$0 = \nabla \cdot \tilde{\mathbf{v}}$$

in Ω and $\tilde{\mathbf{v}} = 0$ on $\partial\Omega$. All eigenvalues λ of (2.13) are either real or occur in complex conjugate pairs. Again, if $\operatorname{Re} \lambda < 0$, the corresponding mode dies out in time. Whereas a mode with $\operatorname{Re} \lambda > 0$ results clearly in instability. Finally, a mode with $\operatorname{Re} \lambda = 0$ is called *neutrally stable* and may trigger nonlinear instability.

If we have countably many eigenvalues with no accumulation point at 0, we conclude that the basic flow of the linearized problem is stable if all normal modes are stable, i.e. $\operatorname{Re} \lambda < 0$ for all eigenvalues. If $\operatorname{Re} \lambda > 0$ for at least one eigenvalue, the basic flow is instable, see [37, 38]. Note that linear instability means in particular nonlinear instability.

Physically, we expect instabilities such as *turbulences* to occur for fast moving inviscid fluids. These characteristics are inherent in a dimensionless quantity, namely the *Reynolds number*. This number is defined as $Re = VL/\nu$ with characteristic velocity V and characteristic length L. Typically, a flow becomes instable as the Reynolds passes a certain threshold, which is called *critical Reynolds number*.

Hence, the critical Reynolds number Re_c is defined as the smallest number such that the basic flow under consideration is stable for all $Re \leq Re_c$, and becomes instable for a $Re > Re_c$. In terms of linear stability theory this means that all eigenvalues of (2.13) have nonnegative real part for $Re \leq Re_c$, and there is at least one eigenvalue with positive real part for a $Re > Re_c$.

2.2.3 The Bénard Problem

We consider a layer of a fluid confined between two parallel planes which is heated from below. If the temperature gradient is sufficiently large to overcome the gravitational force, a tessellated pattern of cellular motion may be observed. This phenomenon is called *Bénard* (or *Rayleigh-Bénard*) *convection*, see [15, 26, 94].

Let the coordinates of the spatial variable $\mathbf{x}$ be denoted by $\mathbf{x} = (x, y, z)$ in the three-dimensional case and by $\mathbf{x} = (x, z)$ in the two-dimensional case. In order to describe the natural convection process, suppose that the fluid is in an infinite layer $z \in (0, l)$, and that we have fixed temperatures T_0 at $z = 0$ and T_l at $z = l$ with $T_0 > T_l$. For the Oberbeck-Boussinesq equations (2.8), there exists a solution at rest with a linear temperature distribution

$$\mathbf{v} = \mathbf{0}, \quad T = -\beta z + T_0,$$

2.2. Linear Stability Theory

where the temperature gradient β is given by

$$\beta = \frac{T_0 - T_l}{d}.$$

The stability of this system is dependent on the dimensionless *Rayleigh number* defined by

$$Ra = \frac{\alpha g \beta}{\kappa \nu} l^4,$$

where g is given by the gravitational vector $\mathbf{g} = (0, 0, -g)^T$.

As for the linear stability analysis, there exists a smallest number Ra_c^{lin} such that for any $Ra > Ra_c^{lin}$ the basic state at rest is linear instable and hence instable. In general, this does not imply that the solution is stable for all $Ra < Ra_c^{lin}$. Considering the nonlinear stability, there exists a critical Rayleigh number Ra_c^{nl} such that the solution is stable for any $Ra < Ra_c^{nl}$. For the solution at rest one can show that $Ra_c^{lin} = Ra_c^{nl}$, see e.g. [94]. In other words, this solution is linear stable if and only if it is nonlinear stable, and that it is linear instable if and only if it is nonlinear instable.

As the Rayleigh number exceeds this critical value, a different steady solution – namely a Bénard convection – can be realized. The critical Rayleigh number for this *primary bifurcation* depends on the boundary setup chosen (rigid or free boundaries) and can for instance be found in [37, 38]. By further increasing the Rayleigh number, more bifurcations in laboratory experiments can be realized, where also time-dependent basic states are observed. Unlike for the primary bifurcation, these depend on the *Prandtl number*

$$Pr = \frac{\nu}{\kappa}$$

as well, see e.g. [37].

In the setup we study later (cf. Section 6.2 and Appendix A) no exact solution of the Oberbeck-Boussinesq equations (2.8) is known. In particular, there exists no solution at rest as for the Bénard problem described above. Therefore, under certain conditions, a simplified model is used for which analytical solutions are known. In Appendix A we evaluate the relative error emerging from this simplification. In this framework, we show the significance of the *Poincaré constant* for these type of evaluations.

The Poincaré constant is an important tool for the qualitative and quantitative description of fluid dynamics in general. Especially, it is essential for the stability behavior as we have seen exemplarily for the stability of the zero solution of the Navier-Stokes equations. In Appendix A we derive a bound for the *Poincaré constant* in a special setup and perform numerical computations which show that the established bound is almost sharp.

2.2.4 Bibliographical Remarks

There are many references on hydrodynamic stability analysis. In addition to the references quoted so far we mention here [26, 37] as books focused on the linear stability theory. The energy method is treated in [94] with an emphasis on thermal convection. Both linear and nonlinear stability are covered in [38, 90]. A relation between linear and nonlinear stability in terms of the Navier-Stokes equations can be found in [89]. Stability of solutions of ordinary differential equations are described in [4, 8]. For a general treatise of stability of solutions in Banach spaces we refer to [31].

2.3 Pseudospectra

As we have seen, eigenvalues are a very important tool to study the stability of a dynamical system. However, in experiments, a fluid flow is observed to become turbulent although an eigenvalue analysis indicates linear stability, see [104] and references therein. This is most notably the case for strongly non normal problems. One remedy is to investigate the nonlinear stability which can assure the stability of a fluid motion. However, to handle nonlinear problems more elaborated techniques than for linear problems are necessary. Since we prefer a linear technique, we consider another encouraging approach by means of pseudospectra, which are a more general tool than spectra.

We consider a dynamical system describing the evolution of a perturbation u by

$$\frac{d}{dt}u = Au,$$

where we assume A to be a linear operator. Under certain conditions, the solution can be expressed in forms of a matrix or operator exponential $e^{tA}u_0$, where u_0 represents the initial disturbance $u(0)$. If the spectrum of A lies in the left half of the complex plane, $e^{tA}u_0$ tends to zero for $t \to \infty$ and no instability is detected. Considering the pseudospectra of A one may discover that $e^{tA}u_0$ becomes arbitrarily large for finite t and hence may trigger instabilities.

In this section we establish lower and upper bounds on e^{tA} by means of pseudospectra. We start by introducing some basic properties of pseudospectra of matrices and linear operators. Our discussion follows [103].

2.3.1 Pseudospectra of Matrices

We begin with three equivalent definitions for pseudospectra of matrices. In what follows we write $(z - A)$ instead of $(zI - A)$, where I is the identity on $\mathbb{C}^n$. Moroever, we use the

2.3. Pseudospectra

convention $\|(z - A)^{-1}\| = \infty$ for any $z \in \sigma(A)$, where $\sigma(A)$ denotes the spectrum of A.

Definition 2.2 *Let $A \in \mathbb{C}^{n \times n}$ and $\varepsilon > 0$. The sets $\sigma_\varepsilon(A)$ defined by*

$$\sigma_\varepsilon(A) = \{z \in \mathbb{C} : \|(z - A)^{-1}\| > \varepsilon^{-1}\}, \tag{2.14}$$

$$\sigma_\varepsilon(A) = \{z \in \mathbb{C} : z \in \sigma(A + E) \text{ for some } E \in \mathbb{C}^{n \times n} \text{ with } \|E\| < \varepsilon\}, \tag{2.15}$$

$$\sigma_\varepsilon(A) = \{z \in \mathbb{C} : \|(z - A)v\| < \varepsilon \text{ for some } v \in \mathbb{C}^n \text{ with } \|v\| = 1\} \tag{2.16}$$

are called ε-pseudospectrum of A.

Note that the definitions depend on the norm. We have chosen definitions using strict inequalities as these lead to equivalent definitions for closed operators in Banach spaces in Section 2.3.2, see [25].

Theorem 2.3 *The above definitions (2.14), (2.15) and (2.16) are equivalent.*

Proof. For $z \in \sigma(A)$ the equivalence is obvious. So we assume $z \notin \sigma(A)$.

To prove (2.15)$\Rightarrow$(2.16), let $(A + E)v = zv$ for some $E \in \mathbb{C}^{n \times n}$ with $\|E\| < \varepsilon$ and $\|v\| = 1$. Then we have $\|(z - A)v\| = \|Ev\| \le \|E\| < \varepsilon$, hence (2.16).

To prove (2.16)$\Rightarrow$(2.14), let $(z - A)v = su$ with $\|u\| = \|v\| = 1$ and $s < \varepsilon$. Thus, we have $\|(z - A)^{-1}\| \ge \|(z - A)^{-1}u\| = \|\frac{1}{s}v\| = \frac{1}{s} > \varepsilon$, which means (2.14).

To prove (2.14)$\Rightarrow$(2.15), suppose $\|(z - A)^{-1}\| > \frac{1}{\varepsilon}$. Then, there exist $s < \varepsilon$ and u, v with $\|u\| = \|v\| = 1$ satisfying $(z - A)^{-1}u = \frac{1}{s}v$. Hence, we have $zv - Av = su$. Next, we show that there exists $E \in \mathbb{C}^{n \times n}$ with $\|E\| = s$ and $Ev = su$ which implies $(A+E)v = zv - su + su = zv$ and hence (2.15). Therefore, we choose E as a rank-1-matrix of the form $E = s u w^H$ with $v^H w = 1$. By virtue of the Hahn-Banach theorem there is a linear functional L on $\mathbb{C}^n$ with $Lv = \|v\| = 1$ and $\|L\| = 1$. Due to the representation theorem of Riesz, there exists a $w \in \mathbb{C}^n$ with $Lv = v^H w$. This implies $|v^H w| = Lv = 1$. From $\|L\| = 1$ we obtain $\max_{\|x\|=1} |x^H w| = 1$. Thus, $\|E\| = s \max_{\|x\|=1} \|u(w^H x)\| = s \max_{\|x\|=1} |w^H x| = s$, which completes the proof. $\square$

The next theorem presents some basic properties of pseudospectra. In this respect we define the sum of two sets C, D by $C + D = \{c + d : c \in C, d \in D\}$.

Theorem 2.4 *Let $A \in \mathbb{C}^{n \times n}$ and $\varepsilon > 0$.*

(1) *$\sigma_\varepsilon(A)$ is nonempty, open, and bounded.*

(2) *If $0 < \varepsilon_1 \le \varepsilon_2$, then $\sigma_{\varepsilon_1}(A) \subseteq \sigma_{\varepsilon_2}(A)$.*

(3) *$\bigcap_{\varepsilon > 0} \sigma_\varepsilon(A) = \sigma(A)$.*

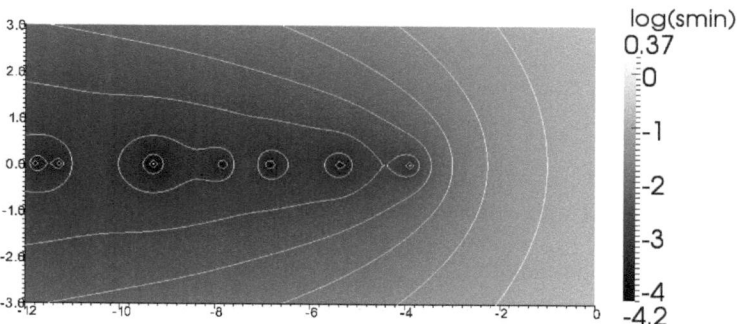

Figure 2.1: Pseudospectra for $\log \varepsilon \in \{0, -0.5, -1, \ldots - 4\}$ with respect to the spectral norm of the discretized convection diffusion operator $-0.05\,\Delta u + 0.3\,u_x + 0.7\,u_y$ defined on a unit square with zero Dirichlet boundary conditions.

(4) $\sigma_\varepsilon(A) \supseteq \sigma(A) + B(\varepsilon)$, where $B(\varepsilon) = \{z \in \mathbb{C} : |z| < \varepsilon\}$ denotes the ε-ball around the origin.

(5) $\sigma_\varepsilon(A)$ consists of at most n connected components, each containing at least one eigenvalue of A.

If we choose the spectral norm, i.e. $\|\cdot\| = \|\cdot\|_2$, we can derive further properties. First, note that the spectral norm of a matrix A corresponds to its largest singular value, i.e. the largest eigenvalue of $A^H A$, see e.g. [98]. This implies the spectral norm of the inverse A^{-1} to be the inverse of the smallest singular value. Thus, by denoting the smallest singular value of $(z - A)$ by $s_{\min}(z - A)$ we have

$$\|(z - A)^{-1}\|_2 = \frac{1}{s_{\min}(z - A)}.$$

This leads to a further representation of pseudospectra, namely,

$$\sigma_\varepsilon(A) = \{z \in \mathbb{C} : s_{\min}(z - A) < \varepsilon\}. \tag{2.17}$$

Theorem 2.5 Let $A \in \mathbb{C}^{n \times n}$ and $\|\cdot\| = \|\cdot\|_2$. Then for all $\varepsilon > 0$,

(1) $\sigma_\varepsilon(A^H) = \overline{\sigma_\varepsilon(A)}$,

(2) $\sigma_\varepsilon(A) = \sigma(A) + B(\varepsilon)$ if and only if A is normal.

Proof. The first assertion follows directly from (2.17). For the second assertion, suppose that A is normal. Then, by the spectral theorem, there exists a unitary matrix U such that

2.3. Pseudospectra

$A = U\Lambda U^H$, where Λ denotes the diagonal matrix with elements equal to the eigenvalues of A. Since for any unitary matrix U we have $\|(z - UAU^H)^{-1}\|_2 = \|U(z - A)^{-1}U^H\|_2 = \|(z - A)^{-1}\|_2$, we conclude

$$\|(z - A)^{-1}\|_2 = \|(z - UAU^H)^{-1}\|_2 = \|(z - \Lambda)^{-1}\|_2$$
$$= \frac{1}{\min_{\lambda_i} |(z - \lambda_i)|} = \frac{1}{\text{dist}(z, \sigma(A))},$$

and hence $\sigma_\varepsilon(A) = \sigma(A) + B(\varepsilon)$. For a proof of the converse statement, we refer to [103]. $\square$

Further properties of pseudospectra can be found in [103]. The close relation between eigenvalues and pseudospectra of matrices is discussed in [40].

2.3.2 Pseudospectra of Linear Operators

The definitions of the last section can be generalized to linear operators acting on an arbitrary Banach space X. We consider closed linear operators $A : D(A) \to X$, where $D(A) \subseteq X$ denotes the domain of A. The set of closed linear operators mapping from X to X is denoted by $\mathcal{C}(X)$. Thus, by definition, $A \in \mathcal{C}(X)$ if for any sequence u_k in $D(A)$ converging to a limit $u \in X$ such that the sequence Au_k is converging to a limit $v \in X$ implies $u \in D(A)$ and $Au = v$. The set of bounded linear operators mapping X into itself is denoted by $\mathcal{B}(X)$. Again we write $(z - A)$ instead of $(zI - A)$, where I denotes the identity on the considered space.

The *spectrum* $\sigma(A)$ of a closed operator is defined by all $z \in \mathbb{C}$ such that either $(z - A)$ is not invertible or $(z - A)^{-1}$ is not bounded on X, see Definition 4.5.

As shown in [103], if $z \notin \sigma(A)$, then $\|(z - A)^{-1}\| \geq \text{dist}(z, \sigma(A))^{-1}$. Therefore, we use the same convention as before by setting $\|(z - A)^{-1}\| = \infty$ for $z \in \sigma(A)$ and proceed with the analogous definitions.

Definition 2.6 *Let $A \in \mathcal{C}(X)$ and $\varepsilon > 0$. The sets $\sigma_\varepsilon(A)$ defined by*

$$\sigma_\varepsilon(A) = \{z \in \mathbb{C} : \|(z - A)^{-1}\| > \varepsilon^{-1}\}, \tag{2.18}$$
$$\sigma_\varepsilon(A) = \{z \in \mathbb{C} : z \in \sigma(A + E) \text{ for some } E \in \mathcal{C}(X) \text{ with } \|E\| < \varepsilon\}, \tag{2.19}$$
$$\sigma_\varepsilon(A) = \{z \in \mathbb{C} : \|(z - A)v\| < \varepsilon \text{ for some } v \in D(A) \text{ with } \|v\| = 1\} \tag{2.20}$$

are called ε-pseudospectrum of A.

Theorem 2.7 *The above definitions (2.18), (2.19) and (2.20) are equivalent.*

Proof. See [103]. □

Theorem 2.8 *Let $A \in \mathcal{C}(X)$ and $\varepsilon > 0$.*

(1) $\sigma_\varepsilon(A)$ *is nonempty and open.*

(2) *If $0 < \varepsilon_1 \leq \varepsilon_2$, then $\sigma_{\varepsilon_1}(A) \subseteq \sigma_{\varepsilon_2}(A)$.*

(3) $\bigcap_{\varepsilon > 0} \sigma_\varepsilon(A) = \sigma(A)$.

(4) $\sigma_\varepsilon(A) \supseteq \sigma(A) + B(\varepsilon)$.

(5) *Any bounded component of $\sigma_\varepsilon(A)$ has a nonempty intersection with $\sigma(A)$.*

As stated in the preceding theorem, some basic properties are similar as in the finite dimensional case (cf. Theorem 2.4). However, in the general case pseudospectra can become unbounded, and only for bounded components it can be assured that they contain elements of the spectrum.

2.3.3 Bounds on Matrix and Operator Exponentials

In studying linear stability, equations of the form

$$\frac{d}{dt} u = Au \qquad (2.21)$$

arise. Here, A is a linear operator acting on a Banach space X. If A is bounded on X, the general solution of (2.21) is given by $u(t) = e^{tA} u_0$, with an initial value u_0 and the exponential function defined by

$$e^{tA} = \sum_{k=0}^{\infty} \frac{1}{k!} t^n A^n. \qquad (2.22)$$

The operator exponential converges for any complex number t. Moreover, it is bounded and analytic with respect to t in the complex plane, see [59]. If A is unbounded, definition (2.22) is not applicable because the domain of A^n may become narrower for increasing n. In this case the operator exponential can be generalized by means of semigroups, see [79].

In this section we present lower and upper bounds for $\|e^{tA}\|$, where we assume A to be a bounded operator on a Banach space X with norm $\|\cdot\|$. Some of these results are also valid for closed operators defining a C_0 semigroup. For this discussion we refer to [79, 103]. In the sequel, we assume X to be a Banach space.

2.3. Pseudospectra

Definition 2.9 *Let $A \in \mathcal{B}(X)$. The* spectral abscissa *of A is defined by*

$$\alpha(A) = \sup_{z \in \sigma(A)} \operatorname{Re} z.$$

Moreover, the ε-pseudospectral abscissa of A is defined by

$$\alpha_\varepsilon(A) = \sup_{z \in \sigma_\varepsilon(A)} \operatorname{Re} z.$$

Theorem 2.10 *Let A be a bounded operator. Then there exist constants $\omega \in \mathbb{R}$ and $M \geq 1$, such that*

$$\|e^{tA}\| \leq M e^{\omega t} \tag{2.23}$$

holds for all $t \geq 0$. For any $z \in \mathbb{C}$ with $\operatorname{Re} z > \omega$ we have $z \notin \sigma(A)$, and, moreover,

$$(z - A)^{-1} = \int_0^\infty e^{-tz} e^{tA} dt, \tag{2.24}$$

and

$$\|(z - A)^{-1}\| \leq \frac{1}{\operatorname{Re} z - \omega}. \tag{2.25}$$

Furthermore, for any closed contour Γ enclosing $\sigma(A)$ in its interior we have

$$e^{tA} = \frac{1}{2\pi i} \int_\Gamma e^{tz} (z - A)^{-1} dz. \tag{2.26}$$

Proof. The inequality (2.23) follows directly from (2.22). For the proofs of (2.24), (2.25) and (2.26) we refer to [79]. □

Theorem 2.11 *Let $\varepsilon > 0$ and L_ε denote the arc length of the boundary of $\sigma_\varepsilon(A)$ or of its convex hull. Then*

$$\|e^{tA}\| \leq \frac{L_\varepsilon e^{t\alpha_\varepsilon(A)}}{2\pi\varepsilon} \tag{2.27}$$

holds for all $t \geq 0$.

Proof. Suppose that $L_\varepsilon < \infty$, otherwise (2.27) obviously holds. Then, by (2.26) we have for any $t \geq 0$ and $\varepsilon > 0$

$$\begin{aligned}\|e^{tA}\| &\leq \frac{1}{2\pi} \int_{\partial\sigma_\varepsilon(A)} |e^{tz}|\, \|(z-A)^{-1}\|\, dz \\ &\leq \frac{1}{2\pi\varepsilon} \int_{\partial\sigma_\varepsilon(A)} e^{t\operatorname{Re} z}\, dz \\ &\leq \frac{1}{2\pi\varepsilon} e^{t\alpha_\varepsilon(A)} L_\varepsilon.\end{aligned}$$

Clearly, these estimates apply for the convex hull of $\sigma_\varepsilon(A)$ as well. □

By replacing $\sigma_\varepsilon(A)$ by its convex hull one may reduce the constant L_ε. Before we state lower bounds dealing with pseudospectra, we present some results involving the spectral abscissa.

Theorem 2.12 *Let $A \in \mathcal{B}(X)$. Then we have*

$$\|e^{tA}\| \geq e^{t\alpha(A)}$$

for all $t \geq 0$ and, moreover, for the strict Lyapunov exponent

$$\lim_{t\to\infty} \frac{1}{t} \log \|e^{tA}\| = \alpha(A).$$

Proof. A proof can be found in [103]. □

Now we are able to state the main result of this chapter which establishes lower bounds on the norm of the operator exponential.

Theorem 2.13 *Let A be a bounded operator acting on a Banach space.*

(1) *The ε-pseudospectral abscissa $\alpha_\varepsilon(A)$ is finite for all $\varepsilon > 0$.*

(2) *If $z \in \mathbb{C}$ with $\operatorname{Re} z > 0$, then*

$$\sup_{t\geq 0} \|e^{tA}\| \geq \operatorname{Re} z\, \|(z-A)^{-1}\|. \tag{2.28}$$

(3) *We have*

$$\sup_{t\geq 0} \|e^{tA}\| \geq \mathfrak{K}(A), \tag{2.29}$$

where

$$\mathfrak{K}(A) := \sup_{\varepsilon \geq 0} \frac{\alpha_\varepsilon(A)}{\varepsilon} = \sup_{\operatorname{Re} z > 0} (\operatorname{Re} z)\|(z-A)^{-1}\|$$

denotes the Kreiss constant.

2.3. Pseudospectra

(4) If $a = \operatorname{Re} z > 0$ and $L = \operatorname{Re} z \, \|(z - A)^{-1}\|$, then

$$\sup_{0 < t \leq \tau} \|e^{tA}\| \geq e^{\tau a} \bigg/ \left(1 + \frac{e^{\tau a} - 1}{L}\right) \tag{2.30}$$

for all $\tau > 0$.

(5) Let $a = \operatorname{Re} z$ and $L = \operatorname{Re} z \, \|(z - A)^{-1}\|$. Suppose that $\|e^{tA}\| \leq M$ for all $t \geq 0$ and $L/M \in (-\infty, 1]$. Then we have

$$\|e^{\tau A}\| \geq e^{\tau a} - \frac{e^{\tau a} - 1}{L/M} = 1 - \frac{(e^{\tau a} - 1)(1 - L'_\tau M)}{L/M}, \tag{2.31}$$

and

$$\|e^{\tau A}\| \geq 1 - \frac{\tau L}{\|(z - A)^{-1}\|}. \tag{2.32}$$

Proof. We set $M = \sup_{t \geq 0} \|e^{tA}\|$ and suppose $M < \infty$. Thus, we can choose $\omega = 0$ in (2.23). By virtue of (2.24) we have for any z with $\operatorname{Re} z > 0$

$$\|(z - A)^{-1}\| \leq M \int_0^\infty e^{t \operatorname{Re} z} dt = \frac{M}{\operatorname{Re} z},$$

which implies (2.28) and hence (2.29).

To prove (2.30), we set $M_\tau = \sup_{0 < t \leq \tau} \|e^{tA}\|$ and suppose $M_\tau < \infty$. Thus, $\|e^{tA}\| \leq M_\tau$ for $0 < t \leq \tau$, $\|e^{tA}\| \leq M_\tau^2$ for $\tau < t \leq 2\tau$, and so on. By (2.24) this implies

$$\|(z - A)^{-1}\| \leq \sum_{j=0}^\infty \int_{j\tau}^{(j+1)\tau} e^{-ta} M_\tau^{j+1} dt = \int_0^\tau e^{-ta} dt \sum_{j=0}^\infty e^{-\tau a j} M_\tau^{j+1},$$

where in the last step t was substituted by $t + j\tau$. If $M_\tau \geq e^{\tau a}$, we have $\tau a > 0$, $L > 1$ and hence (2.30) holds. So we assume $M_\tau < e^{\tau a}$ and we may sum up the geometric series to obtain

$$\|(z - A)^{-1}\| \leq \frac{1}{a}(1 - e^{-\tau a}) \frac{M_\tau}{1 - M_\tau e^{-\tau a}} = \frac{e^{\tau a} - 1}{a(e^{\tau a}/M_\tau - 1)}.$$

Inverting this expression leads to

$$\frac{a}{L} = \frac{1}{\|(z - A)^{-1}\|} \geq \frac{a(e^{\tau a}/M_\tau - 1)}{e^{\tau a} - 1},$$

and consequently

$$\frac{e^{\tau a}}{M_\tau} - 1 \leq \frac{e^{\tau a} - 1}{L},$$

which proves (2.30).

Using (2.30) we show that $\alpha_\varepsilon(A)$ is finite for each $\varepsilon > 0$. Suppose the contrary, that $a = \operatorname{Re} z$ becomes arbitrarily large for some value of $\|(z-A)^{-1}\| = \varepsilon^{-1}$. Setting $\tau = c/a$ in (2.30) for some $c > 0$ yields

$$\sup_{0 < t \le c/a} \|e^{\tau a}\| \ge e^c \bigg/ \left(1 + \frac{(e^c - 1)\varepsilon}{a}\right).$$

Taking $a \to \infty$ shows that $\|e^{\tau a}\|$ must be arbitrarily large for arbitrarily small t, contradicting (2.23).

To prove (2.31) we set $P = \|e^{\tau A}\|$ and for $0 \le t \le \tau$ using (2.23) we conclude

$$\|e^{tA}\| \le M, \quad \|e^{(\tau+t)A}\| \le PM, \quad \|e^{(2\tau+t)A}\| \le P^2 M,$$

and so on. If $P \ge e^{\tau a}$ we have $L/M \ge 1$, $aL \ge 0$ which implies (2.31). For $P < e^{\tau a}$ we get in the same manner as in the proof of (2.30)

$$\frac{L}{a} \le \int_0^\tau e^{-ta} dt \sum_{j=0}^\infty e^{-\tau a j} P^j M = \frac{1}{a}(1 - e^{-\tau a}) \frac{M}{1 - Pe^{-\tau a}},$$

and hence

$$\frac{L}{M} \le \frac{1 - e^{-\tau a}}{1 - Pe^{-\tau a}}.$$

Thus,

$$Pe^{-\tau a} \ge 1 - \frac{1 - e^{-\tau a}}{L/M},$$

which implies (2.31).

Finally taking $a \to 0$ and $L \to 0$ using l'Hôpital's rule proves (2.32). □

For matrices, we can also formulate an upper bound by means of the Kreiss constant, see [103].

Theorem 2.14 (Kreiss Matrix theorem) *For any $A \in \mathbb{C}^{n \times n}$,*

$$\mathfrak{K}(A) \le \sup_{t \ge 0} \|e^{tA}\| \le e \, n \, \mathfrak{K}(A)$$

holds, where $\mathfrak{K}(A)$ is the Kreiss constant of A as defined in Theorem 2.13.

2.3.4 Pseudospectra of Matrix Pencils

As it is shown in Section 4.1, the discretization of an eigenvalue problem by means of finite element methods leads to a generalized matrix eigenvalue problem of the form

$$Au = \lambda Mu. \tag{2.33}$$

One can define the pseudospectrum as $\sigma_\varepsilon(M^{-1}A)$ as in [35]. This may be the natural definition, however, besides assuming M to be nonsingular one might prefer the generalized form for computational reasons. A different definition is established in [84], where the matrix M is assumed to be Hermitian positive definite. In this case the ε-pseudospectrum can be defined by means of the Cholesky factorization $M = F^H F$ as the ε-curve of $\|(z - F^{-H}AF^{-1})^{-1}\|$, where F^{-H} is shorthand for $(F^H)^{-1}$. It is easy to show that assuming $M = F^H F$ and choosing $\|\cdot\| = \|\cdot\|_2$, the latter two definitions are equivalent, see [103].

However, M may be singular as for instance by discretizing the incompressibility constraint of the Navier-Stokes equations. This leads us to the definitions followed in [43], where the pseudospectrum is defined by perturbing both A and M independently.

Definition 2.15 Let $\varepsilon > 0$ and $\alpha, \mu > 0$. Then we define for $A, M \in \mathbb{C}^{n \times n}$ the ε-pseudospectrum of the matrix pencil (2.33) equivalently by

$$\sigma_\varepsilon(A, M) = \{z \in \mathbb{C} : \|(zM - A)^{-1}\| > (\varepsilon(\alpha + |z|\mu))^{-1}\},$$
$$\sigma_\varepsilon(A, M) = \{z \in \mathbb{C} : z \in \sigma(A + A_\delta, M + M_\delta) \text{ for some } A_\delta, M_\delta \in \mathbb{C}^{n \times n} \text{ with}$$
$$\|A_\delta\| < \varepsilon\alpha \text{ and } \|M_\delta\| < \varepsilon\mu\},$$
$$\sigma_\varepsilon(A, M) = \{z \in \mathbb{C} : \|(zM - A)u\| < \varepsilon(\alpha + |z|\mu) \text{ for some } u \in \mathbb{C}^n \text{ with } \|u\| = 1\}.$$

Since we are mainly interested in ε-pseudospectra around the origin, we set $\alpha = 1$ and $\mu = 0$, which is also the definition used in [108, 109]. In this case, Definition 2.15 is applicable if we replace the second identity by

$$\sigma_\varepsilon(A, M) = \{z \in \mathbb{C} : z \in \sigma(A + A_\delta, M) \text{ for some } A_\delta \in \mathbb{C}^{n \times n} \text{ with } \|A_\delta\| < \varepsilon\alpha\}.$$

For nonsingular matrices M, the spectrum of the matrix pencil $\sigma(A, M)$ coincides with the spectrum $\sigma(M^{-1}A)$. However, this property does not hold for pseudospectra in general, but the following inclusions apply

$$\sigma_{\varepsilon/\|M\|}(M^{-1}A) \subset \sigma_\varepsilon(A, M) \subset \sigma_{\varepsilon\|M^{-1}\|}(M^{-1}A). \tag{2.34}$$

From the last inclusion we know that $\sigma_\varepsilon(A, M)$ is bounded for nonsingular M. On the contrary, this is not valid for singular matrices M in general as shown in the next theorem, see [108].

Theorem 2.16 *Let $A, M \in \mathbb{C}^{n \times n}$.*

(1) *If M is nonsingular, then $\sigma_\varepsilon(A, M)$ is bounded for any $\varepsilon > 0$.*

(2) *If M is singular, then $\sigma_\varepsilon(A, M) = \mathbb{C}$ for $\varepsilon > \varepsilon^* = \min\limits_{Mu=0,\, u \neq 0} \dfrac{\|Au\|}{\|u\|}$.*

(3) *If M is not the null matrix, then $\sigma_\varepsilon(A, M) \neq \emptyset$ for all $\varepsilon > 0$.*

Proof. The first statement is a direct consequence of (2.34). The next statement (2) follows from Definition 2.15.

To prove (3), suppose $\sigma(A, M) = \emptyset$ (otherwise, for $z \in \sigma(A, M)$ we have $\|(zM - A)^{-1}\| = \infty$ by convention). Let e_i denote the ith unit vector of $\mathbb{C}^n$. For $1 \leq i, j \leq n$ we define the functions $\varphi_{i,j}(z) = e_i^H (zM - A)^{-1} e_j$, which are analytic in $\mathbb{C}$. As a consequence of Liouville's theorem, each of these functions $\varphi_{i,j}$ are either unbounded or constant. If at least one $\varphi_{i,j}$ is unbounded, then $\|(zM - A)^{-1}\| \geq \varphi_{i,j}(z) \to \infty$ for some $|z| \to \infty$ and therefore $\sigma(A, M)_\varepsilon \neq \emptyset$ for each $\varepsilon > 0$. On the other hand, if all $\varphi_{i,j}$ are constant, then the entries of $(zM - A)^{-1}$ are independent of z, and hence M is the null matrix. □

We have seen in Theorem 2.4 that $\sigma_\varepsilon(A)$ contains the disks with radius ε around the eigenvalues. Furthermore, if A is normal, $\sigma_\varepsilon(A)$ is even identical to the union of these disks. For matrix pencils, this statement can be generalized by

$$\sigma_\varepsilon(A, M) \supseteq \sigma(A, M) + B(\varepsilon/\|M\|),$$

provided that M is not the null matrix. In contrast to pseudospectra of normal matrices, pseudospectra of a pencil (2.33) with A and M normal cannot be determined by its eigenvalues alone. In fact, $\sigma_\varepsilon(A, M)$ can be much larger than the union of disks around $\sigma(A, M)$ with radius $\varepsilon/\|M\|$, see [108].

Chapter 3

Neutron Transport Criticality

The subject of utmost importance for the safety of nuclear reactors is the criticality problem. It examines the evolution of the fission chain reaction of neutrons. If this reaction is in the desirable self-sustained state, the nuclear reactor is said to be critical. In a subcritical state, the population of free neutron decays, whereas in a supercritical state, the chain reaction grows exponentially leading to an uncontrolled explosion. The population of neutrons in a nuclear reactor is modeled by means of the neutron transport equation which finds its applications in medical physics and nuclear radiation shielding technology as well.

In the following Section 3.1 we review the linear Boltzmann equation for modeling neutron transport and explain the basic mechanism of a nuclear reactor. Afterwards, in Section 3.2, we state the criticality problem by means of eigenvalue problems.

3.1 Neutron Transport

The model of neutron transport goes back to Ludwig Boltzmann [17] who established an integro-differential equation for the study of dilute gases. In contrast to the fluid flow model of Section 2.1 it is based on statistical distributions of particles. It has not only been applied successfully on describing dilute gases but also on modeling radiative transport in planetary and stellar atmospheres and neutron transport in nuclear reactors, see [24].

The basic mechanism of the current generation of nuclear reactors (typically using the U^{235} isotope of uranium as fuel) is depicted in Figure 3.1. We consider the three most important types of interactions between neutrons and nuclei. Free neutrons cause *fission reactions* with the nuclei contained in the fuel rods releasing energy as well as fission neutrons. These neutrons are slowed down by the moderator (*scattering reaction*). This enables the chain reaction, because slowed down neutrons are more likely to cause fission

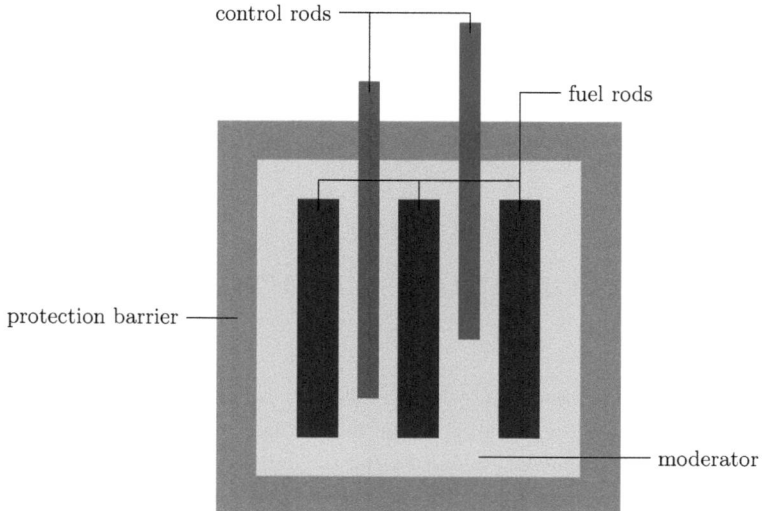

Figure 3.1: Nuclear reactor principle.

reactions. In order for the chain reaction not to run out of control, control rods can be inserted in the nuclear reactor catching free neutrons (*capturing reaction*). By handling these control rods, the nuclear reactor can be kept in a critical state.

We give a brief introduction of the linear Boltzmann transport equation used in reactor physics to understand its basic mechanisms. For a more detailed discussion we refer to [13, 66]. A treatise of nuclear physics with a more introductory character can e.g. be found in [115].

3.1.1 General Assumptions

For the model used in this work, we state the following assumptions. In this context, a *particle* denotes either a neutron or a photon (gamma ray).

(1) Particles may be considered as points, i.e. they can entirely be described by their position and velocity.

(2) Collisions may be considered instantaneous, i.e. after a collision the emerging particles are emitted immediately. For models with delayed neutrons see [13, 66].

(3) Only the expected value of the particle density is taken into account, i.e. fluctuations, which are small in comparison with the average particle density, are neglected.

3.1. Neutron Transport

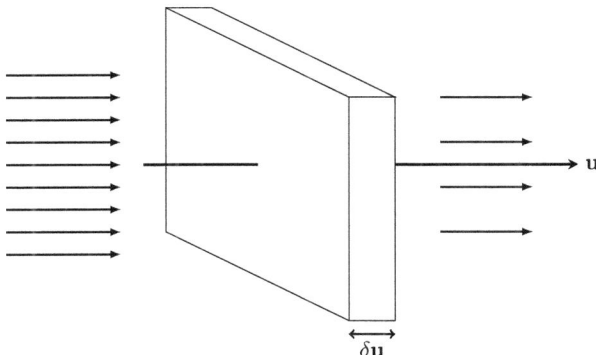

Figure 3.2: A beam of neutrons transmitting through a slab, cf. [66].

(4) Particles travel in straight lines between point collisions.

(5) Since the particle densities in nuclear reactors and other applications are small compared to atomic densities, we may neglect particle-particle interactions.

(6) Material properties are assumed to be isotropic.

For a detailed discussion and physical justifications of these assumptions, we refer to [13, 66].

3.1.2 Cross Section Definitions

At the length scale considered here, we cannot deterministically predict that a particle hits a target nucleus like on a billiard table. These kinds of interactions can only be formulated by means of probabilities which are quantified by *cross sections*.

We consider a beam of particles with intensity I (i.e. I particles per second), with energy E and direction $\mathbf{u}$ impinging perpendicular to a target slab comprised of atoms of a single isotope as in Figure 3.2. The *microscopic* cross section $\tilde{\sigma}(E)$ is the effective cross sectional area per nucleus seen by the particles and is usually measured in *barns* (1 barn $= 10^{-24}$ cm^2). It expresses the probability that a particle interacts with a target nucleus. Let n denote the number of nuclei per unit volume of the medium. Then the beam intensity is governed by

$$I(\mathbf{u} + \delta \mathbf{u}) = I(\mathbf{u})[1 - n\tilde{\sigma}(E)\,\delta \mathbf{u}],$$

where we consider only particles which have not made a collision with nuclei. By

$$\sigma(E) = n\tilde{\sigma}(E)$$

we denote the *macroscopic* cross section, which has units of inverse length. It expresses the probability of particle interaction per unit distance of particle travel. Note that compared to the microscopic cross section, where the probability is based on one target nucleus, the macroscopic cross section quantifies the probability based on one volume unit.

If more than one isotope is considered, the atom densities may not be uniform, and, hence, the macroscopic cross section is in general dependent on the spatial variable **r**. In this case the *total* cross section reads

$$\sigma(\mathbf{r}, E) = \sum_i n^i(\mathbf{r})\, \tilde{\sigma}^i(E),$$

where we use the index i to represent each type of nucleus. The total cross section may be divided into particular cross sections for different types of particle reactions. By denoting the microscopic cross section for a reaction type r for an isotope i as $\tilde{\sigma}_r^i(E)$, we have

$$\sigma_r(\mathbf{r}, E) = \sum_i n^i(\mathbf{r})\, \tilde{\sigma}_r^i(E).$$

We consider *absorption* and *scattering* reactions separately by setting

$$\tilde{\sigma}(E) = \tilde{\sigma}_a(E) + \tilde{\sigma}_s(E).$$

For a discussion of these kinds of cross sections we distinguish between neutron and gamma ray cross sections.

Neutron Cross Sections

In the case of neutron absorption, a sum of various types of physical reactions occur. In nuclear reactor applications, the most important are *capture* and *fission* reactions. In a capture reaction, there is only one capture gamma ray emitted, whereas in a fission reaction a mean value of $\nu(E)$ neutrons and additionally gamma rays are emitted. Thus, we consider

$$\tilde{\sigma}_a(E) = \tilde{\sigma}_c(E) + \tilde{\sigma}_f(E),$$

where $\tilde{\sigma}_c$ is the capture and $\tilde{\sigma}_f$ the fission cross section.

The scattering cross section is comprised of *elastic* $\tilde{\sigma}_n(E)$ and *inelastic* $\tilde{\sigma}_{n'}(E)$ scattering

3.1. Neutron Transport

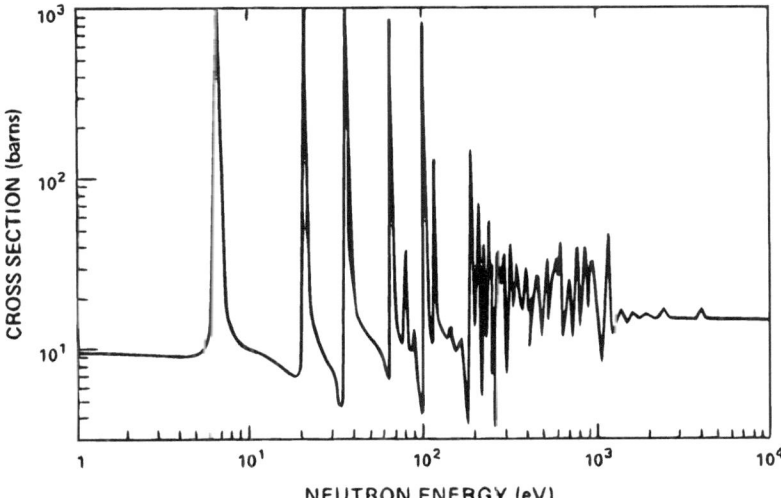

Figure 3.3: Total neutron cross section for U^{238} as a function of incident neutron energy measured in electron volt (1 eV $\approx 1.6 \times 10^{-19}$ joule); adapted from [66].

cross sections:

$$\tilde{\sigma}_s(E) = \tilde{\sigma}_n(E) + \tilde{\sigma}_{n'}(E).$$

In elastic scattering interactions momentum and kinetic energy of the incident particle are conserved, i.e. the impinging particle and the target remain intrinsically unchanged. Whereas in inelastic scattering there is a loss of kinetic energy of the neutron, while the energy state of the nucleus is elevated.

Figure 3.3 shows the complex behavior of neutron cross sections exemplarily for Uranium 238. This data cannot be determined by means of properties of the nuclides. Rather, it has to be determined empirically as a function of energy for each nuclide and for each type of interaction.

Gamma Ray Cross Sections

Unlike neutron cross sections, the significant components of gamma ray cross sections can be determined by first principles or estimated from analytical approximations. We set the absorption cross section equal to the *photoelectric* cross section, which involves absorption of a photon on releasing an electron from one of the orbital shells:

$$\tilde{\sigma}_a(E) = \tilde{\sigma}_{pe}(E).$$

We assume the gamma ray scattering to be comprised of *Compton* scattering and *pair production* scattering:

$$\tilde{\sigma}_s(E) = \tilde{\sigma}_{cs}(E) + \tilde{\sigma}_{pp}(E).$$

Compton scattering consists of scattering of lower energy photons by free electrons. Pair production involves the materialization of a photon into an electron-positron pair, which share the photon energy.

Since gamma ray interactions do not influence the population of neutrons, we can neglect gamma ray cross sections in stating the neutron transport equation in the following section.

3.1.3 The Linear Boltzmann Equation

In the sequel, let Ω denote the neutron direction of travel as it is common notation in neutron transport theory. There should be no confusion with Ω denoting a domain in fluid flow problems.

For a three-dimensional problem we need seven independent variables to describe the distribution of neutrons: three spatial coordinates $\mathbf{r}$, two angles for the neutron direction of travel Ω, the neutron energy E, and the time t. Note that the energy of a neutron can also be expressed by its velocity.

In order to state the neutron transport equation, we first need to introduce some notations. The scattering cross section $\sigma_s(\mathbf{r}, E' \to E, \Omega' \cdot \Omega)$ is defined such that $\sigma_s(\mathbf{r}, E' \to E, \Omega' \cdot \Omega)\,dE\,d\Omega$ quantifies the probability per unit distance of travel that a neutron at position $\mathbf{r}$ with energy E' traveling in direction Ω' will scatter into an energy interval dE about E into a solid angle interval $d\Omega$ about Ω.

Let $\nu(E)$ denote the mean number of neutrons produced in a fission by a neutron with energy E. Furthermore, let $\chi(E)$ be defined such that $\chi(E)\,dE$ is equal to the probability that a neutron produced in a fission will have an energy within an energy interval dE about E.

The neutron *angular density*, denoted by $N(\mathbf{r}, \Omega, E, t)$, is defined as the expected number of neutrons at position $\mathbf{r}$ with direction Ω and energy E at time t per unit volume per unit solid angle per unit energy. The neutron *angular flux* is defined by

$$\psi(\mathbf{r}, \Omega, E, t) = v\,N(\mathbf{r}, \Omega, E, t),$$

where $v = \|\mathbf{v}\|_2$ is the particle speed.

3.1. Neutron Transport

The transport equation expressed in terms of the angular flux ψ reads

$$\left[\frac{1}{v}\frac{\partial}{\partial t} + \boldsymbol{\Omega}\cdot\nabla + \sigma(\mathbf{r},E)\right]\psi(\mathbf{r},\boldsymbol{\Omega},E,t) = q(\mathbf{r},\boldsymbol{\Omega},E,t), \qquad (3.1)$$

where the nabla operator ∇ operates only on the spatial variable $\mathbf{r}$ and q denotes the *emission density*, cf. [13, 66]. The emission density q is comprised of contributions of external sources q_{ex}, scattered neutrons q_s, and fission neutrons q_f, i.e.

$$q = q_{ex} + q_s + q_f.$$

External sources are considered to be known and to be independent of the flux ψ. The emission density for scattered neutrons is given by

$$q_s(\mathbf{r},\boldsymbol{\Omega},E,t) = \int_0^\infty \int_S \sigma_s(\mathbf{r}, E' \to E, \boldsymbol{\Omega}'\cdot\boldsymbol{\Omega})\,\psi(\mathbf{r},\boldsymbol{\Omega}',E',t)\,d\boldsymbol{\Omega}'\,dE',$$

where S denotes the unit sphere. In our case we neglect delayed neutrons, see Section 3.1.1, and thus, the emission density for fission reads

$$q_f(\mathbf{r},\boldsymbol{\Omega},E,t) = \chi(E)\int_0^\infty \int_S \nu(E')\,\sigma_f(\mathbf{r},E')\psi(\mathbf{r},\boldsymbol{\Omega}',E',t)\,d\boldsymbol{\Omega}'\,dE'.$$

Inserting q in equation (3.1) implies

$$\begin{aligned}\left[\frac{1}{v}\frac{\partial}{\partial t} + \boldsymbol{\Omega}\cdot\nabla + \sigma(\mathbf{r},E)\right]\psi(\mathbf{r},\boldsymbol{\Omega},E,t) &= q_{ex}(\mathbf{r},\boldsymbol{\Omega},E,t) \\ &+ \int_0^\infty \int_S \sigma_s(\mathbf{r},E'\to E, \boldsymbol{\Omega}'\cdot\boldsymbol{\Omega})\,\psi(\mathbf{r},\boldsymbol{\Omega}',E',t)\,d\boldsymbol{\Omega}'\,dE' \qquad (3.2)\\ &+ \chi(E)\int_0^\infty \int_S \nu(E')\,\sigma_f(\mathbf{r},E')\,\psi(\mathbf{r},\boldsymbol{\Omega}',E',t)\,d\boldsymbol{\Omega}'\,dE'.\end{aligned}$$

Let V denote the domain within which the neutron transport problem is to be solved and let $\Gamma = \partial V$ be its boundary. To solve the transport equation (3.2), one has to impose an initial condition $\psi(\mathbf{r},\boldsymbol{\Omega},E,0)$ at $t = 0$ as well as appropriate boundary conditions. Let $\mathbf{n}$ denote the unit vector normal to Γ pointing outwards. Then the incoming flux is specified by

$$\psi(\mathbf{r},\boldsymbol{\Omega},E,t) = \psi_{in}(\mathbf{r},\boldsymbol{\Omega},E,t), \quad \mathbf{n}\cdot\boldsymbol{\Omega} < 0, \quad \mathbf{r}\in\Gamma.$$

The common case where $\psi_{in} = 0$ is referred to as *vacuum boundary condition*. *Reflective boundary conditions*, which prescribe that all outgoing neutrons are reflected back, are characterized by

$$\psi(\mathbf{r}, \mathbf{\Omega}, E, t) = \psi(\mathbf{r}, \mathbf{\Omega}', E, t), \quad \mathbf{n} \cdot \mathbf{\Omega} = -\mathbf{n} \cdot \mathbf{\Omega}', \quad (\mathbf{\Omega} \times \mathbf{\Omega}') \cdot \mathbf{n} = 0, \quad \mathbf{r} \in \Gamma,$$

where $\mathbf{\Omega}'$ denotes the incident direction and $\mathbf{\Omega}$ the reflection direction.

3.2 Criticality

The stability of a system containing fissile nuclides is characterized by its population of free neutrons. If the number of free neutron is decaying in time, which means the fission reaction dies out in time, the system is said to be *subcritical*. On the other hand, it is called *supercritical* if the population of free neutrons is growing. Finally, it is defined to be *critical* if the number of free neutron reaches a time independent equilibrium in the absence of external sources of neutrons, i.e. if there exists a steady nonnegative solution of the source free transport equation (3.2)

$$\left[\mathbf{\Omega} \cdot \nabla + \sigma(\mathbf{r}, E)\right] \psi(\mathbf{r}, \mathbf{\Omega}, E) = \int_0^\infty \int_S \sigma_s(\mathbf{r}, E' \to E, \mathbf{\Omega}' \cdot \mathbf{\Omega}) \psi(\mathbf{r}, \mathbf{\Omega}', E') \, d\mathbf{\Omega}' \, dE' \\ + \chi(E) \int_0^\infty \int_S \nu(E') \sigma_f(\mathbf{r}, E') \psi(\mathbf{r}, \mathbf{\Omega}', E') \, d\mathbf{\Omega}' \, dE', \quad (3.3)$$

with appropriate boundary conditions (e.g. vacuum or reflective boundary conditions). These criticality characterizations can be reformulated as eigenvalue problems.

3.2.1 The α Eigenvalue

Suppose that we have asymptotic solutions of the form

$$\psi(\mathbf{r}, \mathbf{\Omega}, E, t) = \psi_\alpha(\mathbf{r}, \mathbf{\Omega}, E) \, e^{\alpha t}$$

satisfying the imposed initial and boundary conditions. Inserting this to the source free formulation of (3.2) yields the eigenvalue problem

$$\left[\frac{\alpha}{v} + \mathbf{\Omega} \cdot \nabla + \sigma(\mathbf{r}, E)\right] \psi_\alpha(\mathbf{r}, \mathbf{\Omega}, E) = \int_0^\infty \int_S \sigma_s(\mathbf{r}, E' \to E, \mathbf{\Omega}' \cdot \mathbf{\Omega}) \psi_\alpha(\mathbf{r}, \mathbf{\Omega}', E') \, d\mathbf{\Omega}' \, dE' \\ + \chi(E) \int_0^\infty \int_S \nu(E') \sigma_f(\mathbf{r}, E') \psi_\alpha(\mathbf{r}, \mathbf{\Omega}', E') \, d\mathbf{\Omega}' \, dE'.$$

Assume that there exists an expansion of the solution ψ_α in eigenfunctions ψ_i. Let α_0 denote the eigenvalue having the largest real part and ψ_0 an associated eigenfunction. For large t, we expect that the solution of the initial value problem is proportional to

3.2. Criticality

$\psi_0\,e^{\alpha_0 t}$. Furthermore, we assume for physical reasons that α_0 is real, otherwise negative or imaginary densities could occur. Thus, we make the following distinction by the sign of α_0 for characterizing the criticality:

$$\alpha_0 \begin{cases} > 0: & \text{supercritical,} \\ = 0: & \text{critical,} \\ < 0: & \text{subcritical.} \end{cases}$$

3.2.2 The k Eigenvalue

To derive the k eigenvalue form, we assume that the system can be made critical by adjusting the number of neutrons emitted by fission. That means we can replace ν by ν/k in (3.3) resulting in the eigenvalue problem

$$\begin{aligned}\left[\mathbf{\Omega}\cdot\nabla + \sigma(\mathbf{r},E)\right]\psi(\mathbf{r},\mathbf{\Omega},E) =& \\ &\int_0^\infty \int_S \sigma_s(\mathbf{r},E'\to E,\mathbf{\Omega}'\cdot\mathbf{\Omega})\,\psi(\mathbf{r},\mathbf{\Omega}',E')\,d\mathbf{\Omega}'\,dE' \\ +& \frac{\chi(E)}{k}\int_0^\infty \int_S \nu(E')\,\sigma_f(\mathbf{r},E')\,\psi(\mathbf{r},\mathbf{\Omega}',E')\,d\mathbf{\Omega}'\,dE'.\end{aligned} \quad (3.4)$$

If the largest eigenvalue k is equal to one, the criticality condition (3.3) is clearly satisfied. The case $k < 1$ means that the number of neutrons per fission to make the system critical is larger than the actual ν, i.e. the system is subcritical. Conversely, it is supercritical for $k > 1$. Wrapping up, we have:

$$k \begin{cases} > 1: & \text{supercritical,} \\ = 1: & \text{critical,} \\ < 1: & \text{subcritical.} \end{cases}$$

If the system is critical, i.e. $\alpha_0 = 0$ and $k = 1$, the corresponding eigenfunctions are identical. However, for any subcritical or supercritical system, the eigenfunctions may differ. In the α eigenvalue problem we have an additional term α_0/v which is referred to as *time absorption*. Since this term may cause difficulties in numerical computations, one treats the criticality problem by evaluating the k eigenvalue rather than the α eigenvalue in many applications, see [13, 66].

Chapter 4

Discretization of the Eigenvalue Problems

In order to solve the eigenvalue problems arising in the context of hydrodynamic stability (2.13) and criticality (3.4) numerically, we need appropriate discretization techniques. In the following Section 4.1 we derive a priori error estimates for the spectrum and the pseudospectrum in terms of a finite element approximation. Afterwards, in Section 4.2 we treat the discretization of the k eigenvalue criticality problem.

4.1 Galerkin Finite Element Spectral Approximation

The spectral approximation theory we present is put in the framework of elliptic differential operators which arise in the field of fluid dynamics. It is formulated by means of a bounded operator with a compact inverse in order to apply the approximation theory of compact operators. This presentation follows [16, 53].

4.1.1 Problem Formulation for Elliptic Eigenvalue Problems

Variational Formulation

Let $\mathcal{A}$ be a linear differential operator defined on a bounded domain $\Omega \subset \mathbb{R}^d$ of order $2m$:

$$\mathcal{A}u = \sum_{|\alpha|\leq m} \sum_{|\beta|\leq m} (-1)^{|\beta|} \partial^\beta a_{\alpha\beta}(x) \partial^\alpha u. \tag{4.1}$$

We assume the operator $\mathcal{A}$ to be *uniformly elliptic*, i.e. there exists a constant $\varepsilon > 0$ such that

$$\sum_{|\alpha|,|\beta|=m} a_{\alpha\beta}(x)\xi^{\alpha+\beta} \geq \varepsilon |\xi|^{2m} \quad \forall x \in \Omega, \forall \xi \in \mathbb{R}^d.$$

The *classical formulation* of the eigenvalue problem for this operator reads

$$\begin{aligned} \text{Find } \lambda \in \mathbb{C},\ u \neq 0 \text{ such that } \mathcal{A}u &= \lambda u \quad \text{in } \Omega, \\ B_j u &= 0 \quad \text{on } \partial\Omega, \quad j = 1, \ldots, m, \end{aligned} \quad (4.2)$$

where B_j denote appropriate boundary operators. Note that inhomogeneous Dirichlet boundary conditions are not allowed since we need the space of functions fulfilling the boundary conditions to be a vector space.

The associated *variational formulation* (or *weak formulation*) of (4.2) is given by:

$$\text{Find } \lambda \in \mathbb{C},\ u \in V \text{ such that } a(u,\varphi) = \lambda\,(u,\varphi)_0 \quad \forall \varphi \in V, \quad \|u\|_0 = 1, \quad (4.3)$$

where $a : V \times V \to \mathbb{C}$ is the sesquilinear form generated by the operator $\mathcal{A}$. We assume V to be an appropriate (complex) Sobolev space such that $V \subset L^2(\Omega) \subset V'$ builds a Gelfand triple, i.e. $V \subset L^2(\Omega)$ is continuously and densely embedded (for instance $V = H_0^m(\Omega)$). Here, V' denotes the continuous dual space of V.

In the case of homogeneous Dirichlet boundary conditions, the corresponding sesquilinear form reads

$$a(u,\varphi) = \int_\Omega \sum_{|\alpha|,|\beta| \leq m} a_{\alpha\beta}(x)\, \partial^\alpha u\, \overline{\partial^\beta \varphi}\, d\mu. \quad (4.4)$$

The adjoint eigenvalue problem associated to (4.3) seeks $\lambda^* \in \mathbb{C}$ and $u^* \in V$ such that

$$a(\varphi, u^*) = \lambda^*\,(u^*, \varphi)_0 \quad \forall \varphi \in V, \quad \|u^*\|_0 = 1. \quad (4.5)$$

Clearly, the primal and adjoint eigenvalues are related to each other by $\lambda^* = \overline{\lambda}$.

Further, we assume the sesquilinear form $a(\cdot,\cdot)$ to be *V-coercive*. This means it is *bounded* (or *continuous*), i.e.

$$|a(u,v)| \leq C_B \|u\|_V \|v\|_V \quad \forall u,v \in V,$$

with a constant $C_B > 0$, and there exist $C_K \in \mathbb{R}$ and $C_E > 0$ such that

$$\operatorname{Re} a(v,v) \geq C_E \|v\|_V^2 - C_K \|v\|_0^2 \quad \forall v \in V. \quad (4.6)$$

For $V = H_0^m(\Omega)$, assuming sufficiently smooth coefficients $a_{\alpha\beta}$, this coercivity property is derived by the uniform ellipticity of $\mathcal{A}$ using Gårding's theorem, see e.g. [21, 120].

The sesquilinear form $\tilde{a} : V \times V \to \mathbb{C}$ defined by $\tilde{a}(u,v) = a(u,v) + C_K(u,v)_0$ possesses the same eigenfunctions as $a(\cdot,\cdot)$, and the eigenvalues of $\tilde{a}(\cdot,\cdot)$ are the same eigenvalues of $a(\cdot,\cdot)$ shifted by the constant C_K. Therefore, without loss of generality, we can assume

4.1. Galerkin Finite Element Spectral Approximation

$C_K = 0$ in (4.6), i.e. $a(\cdot,\cdot)$ is V-elliptic.

Furthermore, we assume the embedding $V \subset L^2(\Omega)$ to be compact. For $V = H_0^m(\Omega)$ or $V = H^m(\Omega)$ and assuming the boundary $\partial\Omega$ to be sufficiently smooth, this is a consequence of the Rellich-Kondrachov compactness theorem, see [3, 116].

Let $A \in \mathcal{B}(V, V')$ be the unique linear operator associated to the sesquilinear form $a(\cdot,\cdot)$, i.e. $(Au)(v) = a(u,v)$. Let $\mathcal{B}(X,Y)$ denote the space of bounded linear operators from X to Y. Due to the V-ellipticity of $a(\cdot,\cdot)$, we have that $A^{-1} \in \mathcal{B}(V', V)$.

Definition 4.1 *With the continuous, dense and, compact embeddings $I_{V \hookrightarrow L^2(\Omega)}$, $I_{L^2(\Omega) \hookrightarrow V'}$ and $I_{V \hookrightarrow V'}$, we define the operators*

$$T : L^2(\Omega) \to L^2(\Omega), \qquad T = I_{V \hookrightarrow L^2(\Omega)} \circ A^{-1} \circ I_{L^2(\Omega) \hookrightarrow V'},$$
$$\tilde{T} : V \to V, \qquad \tilde{T} = A^{-1} \circ I_{V \hookrightarrow V'}.$$

Lemma 4.2 *The operators T and $\tilde{T}$ are compact.*

Proof. T and $\tilde{T}$ are defined as compositions of compact and bounded linear operators and are therefore compact. □

Note that λ is an eigenvalue of the variational formulation (4.3) if and only if $\mu = \frac{1}{\lambda}$ is an eigenvalue of T. This holds for $\tilde{T}$ respectively.

Galerkin Finite Element Discretization

Let $V_h \subset V$ be a finite-dimensional subspace endowed with the norm $\|\cdot\|_V$. The approximations of the eigenvalue problems (4.3) and (4.5) by a Galerkin finite element method read:

Find $\lambda_h \in \mathbb{C}$, $u_h \in V_h$ such that $a(u_h, \varphi_h) = \lambda_h (u_h, \varphi_h)_0 \quad \forall \varphi_h \in V_h, \quad \|u_h\|_0 = 1.$ (4.7)

Find $\lambda_h^* \in \mathbb{C}$, $u_h^* \in V_h$ such that $a(\varphi_h, u_h^*) = \lambda_h^* (u_h^*, \varphi_h)_0 \quad \forall \varphi_h \in V_h, \quad \|u_h^*\|_0 = 1.$ (4.8)

Again, we have that $\lambda_h^* = \overline{\lambda}_h$. In algebraic notation this results in the following generalized eigenvalue problems

$$\mathbf{A}_h x_h = \lambda_h \mathbf{M}_h x_h, \qquad (4.9)$$
$$\mathbf{A}_h^H x_h^* = \lambda_h^* \mathbf{M}_h x_h^*, \qquad (4.10)$$

where $\mathbf{A}_h = \left(a(\varphi_j^h, \varphi_i^h)\right)_{i,j}$ is the *stiffness matrix* and $\mathbf{M}_h = \left((\varphi_h^j, \varphi_h^i)_0\right)_{i,j}$ is the symmetric and positive definite *mass matrix*. Here $\{\varphi_i^h\}$ denotes a basis of V_h. If we have real coefficients $a_{\alpha\beta}$ and choose the basis $\{\varphi_i\}$ to be real, we obtain real matrices $\mathbf{A}_h$ and $\mathbf{M}_h$.

Analogous to the continuous case, we define $A_h^{-1} : V' \to V_h$ as the bounded linear operator such that for $f \in V'$ we have $A_h^{-1} f = u_h$, where u_h is the solution of

$$a(u_h, \varphi_h) = (f, \varphi_h)_0 \quad \forall \varphi_h \in V_h.$$

Now we are able to define the discrete counterpart of T and $\tilde{T}$.

Definition 4.3 *With the embeddings $I_{V_h \hookrightarrow L^2(\Omega)}$, $I_{L^2(\Omega) \hookrightarrow V'}$, $I_{V_h \hookrightarrow V}$ and $I_{V \hookrightarrow V'}$, we define the operators*

$$T_h : L^2(\Omega) \to L^2(\Omega), \quad T = I_{V_h \hookrightarrow L^2(\Omega)} \circ A_h^{-1} \circ I_{L^2(\Omega) \hookrightarrow V'},$$
$$\tilde{T}_h : V \to V, \quad \tilde{T} = I_{V_h \hookrightarrow V} \circ A^{-1} \circ I_{V \hookrightarrow V'}.$$

Lemma 4.4 *The operators T_h and $\tilde{T}_h$ are compact.*

Proof. T_h and $\tilde{T}_h$ are defined as compositions of compact and bounded linear operators and are therefore compact. $\square$

Again, note that λ_h is an eigenvalue solution of the discrete formulation (4.7) if and only if $\mu_h = \frac{1}{\lambda_h}$ is an eigenvalue of T_h. This holds for $\tilde{T}_h$ respectively.

4.1.2 Spectral Approximation of Compact Operators

The previously defined differential operators A and A_h have compact inverses. Therefore, we can apply the spectral approximation theory of compact operators based on [75]. We start with some main results about compact operators including the Riesz-Schauder theory, see e.g. [59, 113, 120].

In the sequel, we write $(z - T)$ instead of $(zI - T)$, where I denotes the identity. Throughout this section, let X be a Banach space.

Definition 4.5 *Let $T \in \mathcal{C}(X)$ be a closed linear operator.*

(1) *The* resolvent set $\rho(T)$ *of T is defined by*

$$\rho(T) = \{z \in \mathbb{C} : (z - T)^{-1} \text{ exists in } \mathcal{B}(X)\}.$$

(2) *For any $z \in \rho(T)$*

$$R_z = R_z(T) = (z - T)^{-1}$$

defines the resolvent *of T.*

4.1. Galerkin Finite Element Spectral Approximation

(3) The set
$$\sigma(T) = \mathbb{C} \setminus \rho(T)$$
is called the spectrum of T.

Any $\mu \in \sigma(T)$ with $N(\mu - T) \neq \{0\}$ is called eigenvalue of T, where N denotes the kernel. For an eigenvalue μ, the set $N(\mu - T)$ is called eigenspace of T associated to μ, and any element $u \in N(\mu - T)$, $u \neq 0$ is called eigenvector or eigenfunction of T.

Theorem 4.6 (Riesz-Schauder theory) Let $T : X \to X$ be a compact linear operator.

(1) The set $\sigma(T)\setminus\{0\}$ consists of at most countably many elements.

(2) Each $\mu \in \sigma(T)\setminus\{0\}$ is an eigenvalue of T.

(3) For any $\varepsilon > 0$ the set $\{\lambda \in \sigma(T) : |\lambda| \geq \varepsilon\}$ is finite, i.e. the spectrum $\sigma(T)$ possesses at most one accumulation point at 0.

(4) For any $\mu \in \sigma(T)\setminus\{0\}$ the ascent α_μ defined by

$$\alpha_\mu = \max\{\alpha \in \mathbb{N} : N\left((\mu - T)^{\alpha-1}\right) \neq N\left((\mu - T)^\alpha\right)\}$$

is finite. Moreover, the geometric multiplicity defined by $\dim N(\mu - T)$ and the algebraic multiplicity defined by $\dim N((\mu - T)^{\alpha_\mu})$ are finite. For $\mu \in \sigma(T)\setminus\{0\}$, the elements of $N((\mu-T)^{\alpha_\mu})$ are called the generalized eigenfunctions of T associated to μ.

(5) For each $\mu \in \sigma(T)\setminus\{0\}$ we have the direct decomposition

$$X = N\left((\mu - T)^{\alpha_\mu}\right) \oplus R\left((\mu - T)^{\alpha_\mu}\right),$$

where R denotes the range.

(6) If $\mu \in \sigma(T)\setminus\{0\}$, then the identity $\sigma(T|_{R((\mu-T)^{\alpha_\mu})}) = \sigma(T)\setminus\{\mu\}$ holds.

Theorem 4.7 Let $T \in \mathcal{C}(X)$.

(1) The resolvent set $\rho(T)$ is open in $\mathbb{C}$.

(2) The function $\mathbb{C} \ni z \mapsto R_z(T) \in \mathcal{B}(X)$ is analytic in each connected component of $\rho(T)$.

(3) If T is compact, then in the neighborhood of an eigenvalue μ with ascent α_μ the resolvent can be expanded in a Laurent series:

$$R_z(T) = \sum_{k=-\alpha_\mu}^{\infty} A_k(z-\mu)^k, \quad \text{where } A_k = \frac{1}{2\pi i} \int_\Gamma \frac{R_\zeta(T)}{(\zeta-\mu)^{(k+1)}} d\zeta,$$

and Γ is a Jordan curve enclosing μ and containing or intersecting no other elements of $\sigma(T)$.

Definition 4.8 Let $T \in \mathcal{B}(X)$ be compact. Then for $0 \neq \mu \in \sigma(T)$ the Dunford integral given by

$$E(\mu, T) = \frac{1}{2\pi i} \int_\Gamma R_z(T) \, dz,$$

where Γ is a Jordan curve enclosing μ and not containing or intersecting any other elements of $\sigma(T)$, is called spectral projector of T. By Cauchy's integral theorem, this definition does not depend on the choice of the Jordan curve Γ.

Note that $E(\mu, T)$ corresponds to the residue of $R_z(T)$, i.e. $E(\mu, T) = A_{-1}$. In the following theorem we present some properties of the spectral projector which justify its denotation.

Theorem 4.9 Let $T \in \mathcal{B}(X)$ be compact.

(1) If $\mu \in \sigma(T)\setminus\{0\}$, then $E(\mu, T)$ is a projection.

(2) If $\mu_1, \mu_2 \in \sigma(T)\setminus\{0\}$ and $\mu_1 \neq \mu_2$, then $E(\mu_1, T)E(\mu_2, T) = 0$.

(3) Let α_μ denote the ascent of the eigenvalue $\mu \in \sigma(T)\setminus\{0\}$. Then the range $R(E(\mu, T))$ of the spectral projector corresponds to the space of generalized eigenfunctions associated to μ, i.e.

$$R(E(\mu, T)) = N((\mu - T)^{\alpha_\mu}).$$

In the following lemma we show that an approximation of a bounded linear operator approximates the spectrum and the resolvent as well.

Lemma 4.10 Let $F \in \mathcal{B}(X)$ and $\varepsilon > 0$. Then there exists $\delta > 0$ such that for all $S \in \mathcal{B}(X)$ with $\|F - S\| < \delta$ we have

$$\sigma(S) \subset B_\varepsilon(\sigma(F)),$$
$$\|R_z(F) - R_z(S)\| < \varepsilon \quad \text{for any } z \in B_\varepsilon(\sigma(F)),$$

4.1. Galerkin Finite Element Spectral Approximation

where $B_\varepsilon(\sigma(F))$ denotes the ε-disk around $\sigma(F)$, i.e.

$$B_\varepsilon(\sigma(F)) = \{\mu + \xi : \mu \in \sigma(F), |\xi| < \varepsilon\}.$$

Proof. i) First, we show that the mapping $A \mapsto A^{-1}$ is continuous for invertible bounded linear operators. Assume $A \in \mathcal{B}(X)$ is invertible. The set of invertible operators in $\mathcal{B}(X)$ is open. Any $B \in \mathcal{B}(X)$ with $\|B - A\|\|A^{-1}\| < 1$ is invertible and we have

$$B^{-1} = (A - (B - A))^{-1} = A^{-1}(I - (B-A)A^{-1})^{-1} = A^{-1}\sum_{k=0}^{\infty}\left[(B-A)A^{-1}\right]^k.$$

Thus, we obtain

$$\|A^{-1} - B^{-1}\| \le \|A^{-1}\|\sum_{k=1}^{\infty}(\|B-A\|\|A^{-1}\|)^k = \frac{\|B-A\|\|A^{-1}\|^2}{1 - \|B-A\|\|A^{-1}\|} \to 0 \quad (B \to A),$$

which implies the continuity of the inverse mapping for linear bounded operators.

ii) Since $(I - z^{-1}F) \to I$ as $z \to \infty$ and using i) we obtain

$$\lim_{z \to \infty}\|R_z(F)\| = \lim_{z \to \infty}\|\frac{1}{z}(I - z^{-1}F)\| = 0.$$

Hence, there exists $N_\varepsilon \ge 0$ satisfying $\|R_z(F)\| \le N_\varepsilon$ for any $z \in \mathbb{C}\setminus B_\varepsilon(\sigma(F))$.

iii) Let $z \in \mathbb{C}\setminus B_\varepsilon(\sigma(F))$ and $\|S - F\| < \delta := N_\varepsilon^{-1}$. Then we have

$$R_z(F)\sum_{i=0}^{\infty}[(S-F)R_z(F)]^i = (z-F)^{-1}\left\{(I - (S-F)(I-F)^{-1}\right\}^{-1} \qquad (4.11)$$
$$= (z-S)^{-1} = R_z(S),$$

which implies that $(z - S)^{-1}$ exists and is bounded and therefore $z \in \rho(S)$ holds. Consequently for $\|S - F\| < \tilde{\varepsilon}_1$ we have $\sigma(S) \subset B_\varepsilon(\sigma(F))$.

iv) Let $\|S - F\| \le \tilde{\delta} := \varepsilon\left(N_\varepsilon^2 + \varepsilon N_\varepsilon\right)^{-1} < N_\varepsilon^{-1}$. Using (4.11) we have

$$\|R_z(S) - R_z(F)\| \le \|R_z(F)\|\sum_{i=1}^{\infty}\|(S-F)R_z(F)\|^i \le \frac{N_\varepsilon^2\|S - F\|}{1 - \|S - F\|N_\varepsilon} < \varepsilon,$$

which completes the proof. $\square$

In the sequel, let T and $\{T_h\}_{h>0}$ be compact operators defined on a Hilbert space such that

$$\lim_{h \to 0}\|T - T_h\| = 0. \qquad (4.12)$$

This assertion is satisfied in the context of finite elements and will be proven later (Section 4.1.3). We proceed by deriving a priori convergence estimates for eigenpairs of compact operators.

Let Γ be a Jordan curve enclosing an eigenvalue $\mu \in \sigma(T)\setminus\{0\}$ as in Definition 4.8. If $\{T_h\}_{h>0}$ is a sequence of compact operators satisfying (4.12), Lemma 4.10 shows that for h small enough we have $\Gamma \subset \rho(T_h)$, and therefore, analogously to Definition 4.8, we can define the operator

$$E(\mu, T_h) = \frac{1}{2\pi i} \int_\Gamma R_z(T_h)\, dz.$$

For brevity, we set $E = E(\mu, T)$ and $E_h = E(\mu, T_h)$ if the arguments used are apparent from the context.

Lemma 4.11 *The operator $E(\mu, T_h)$ is a spectral projector onto the direct sum of the generalized eigenfunction spaces corresponding to the eigenvalues of T_h enclosed by Γ.*

Lemma 4.12 *For $\mu \in \sigma(T)$ we have*

$$\|E(\mu, T) - E(\mu, T_h)\| \to 0, \quad \text{as } h \to 0.$$

The following convergence result is stronger than Lemma 4.10 since it assures the convergence of eigenvalues according to their algebraic multiplicity.

Theorem 4.13 *For an eigenvalue $\mu \in \sigma(T)$ let σ_μ denote its algebraic multiplicity. Given a Jordan curve Γ enclosing μ and excluding all other eigenvalues of T, there exists $h_0 > 0$ such that for all $0 < h < h_0$ the curve Γ encloses exactly σ_μ eigenvalues of T_h counted with their algebraic multiplicities.*

The considered eigenvalues of T_h in the preceding Theorem 4.13 which lie inside Γ are denoted by $\mu_1(h), \ldots \mu_{\sigma_\mu}(h)$.

Definition 4.14 *Let M, N be two subspaces of a Hilbert space X. We define*

$$\delta_X(M, N) = \sup_{x \in M, \|x\|=1} \text{dist}(x, N),$$

and the gap $\hat{\delta}_X(M, N)$ between M and N as

$$\hat{\delta}_X(M, N) = \max(\delta(M, N), \delta(N, M)).$$

We write $\delta(M, N)$ and $\hat{\delta}(M, N)$ if there is no confusion about the underlying Hilbert space X.

4.1. Galerkin Finite Element Spectral Approximation

Theorem 4.15 *For h small enough, there is a constant C independent of h such that*

$$\hat{\delta}(R(E), \bar{R}(E_h)) \leq \|(E - E_h)|_{R(E)}\| \leq C\|(T - T_h)|_{R(E)}\|.$$

Proof. See [75, Theorem 1]. □

Theorem 4.13 shows that all eigenvalues $\mu_1(h), \ldots, \mu_{\sigma_\mu}(h)$ converge to μ. However, the individual $\mu_i(h)$ might be rather poor approximations to μ. Nevertheless their arithmetic mean is generally a better approximation, cf. [20]. Thus, we define

$$\hat{\mu}(h) = \frac{1}{\sigma_\mu} \sum_{i=1}^{\sigma_\mu} \mu_i(h).$$

Theorem 4.16 *For h small enough, there is a constant C independent of h such that*

$$|\mu - \hat{\mu}(h)| \leq C\|(T - T_h)|_{R(E)}\|.$$

Proof. See [75, Theorem 2]. □

A refined estimation is given in the following theorem.

Theorem 4.17 *Let $\varphi_1, \ldots, \varphi_{\sigma_\mu}$ be an orthonormal basis of $R(E)$ and $\varphi_k^* = E^* \varphi_k$ for $k = 1, \ldots, \sigma_\mu$. For h small enough, there is a constant C independent of h such that*

$$|\mu - \hat{\mu}(h)| \leq \frac{1}{\sigma_\mu} \sum_{j=1}^{\sigma_\mu} |((T - T_h)\varphi_j, \varphi_j^*)| + C \, \|(T - T_h)|_{R(E)}\| \, \|(T^* - T_h^*)|_{R(E^*)}\|.$$

Now we state the convergence of separate eigenvalues $\mu_k(h)$ instead of their arithmetic mean. In this view, the order of convergence decreases by means of the ascent α_μ of μ.

Theorem 4.18 *Let $\varphi_1, \ldots, \varphi_{\sigma_\mu}$ be an orthonormal basis of $R(E)$ and $\varphi_k^* = E^* \varphi_k$ for $k = 1, \ldots, \sigma_\mu$. For h small enough, there is a constant C independent of h such that*

$$|\mu - \mu_k(h)|^{\alpha_\mu} \leq C \left\{ \sum_{i,j=1}^{\sigma_\mu} |((T - T_h)\varphi_i, \varphi_j^*)| + \|(T - T_h)|_{R(E)}\| \, \|(T^* - T_h^*)|_{R(E^*)}\| \right\}.$$

So far, we have only stated that eigenvectors of the discrete problem approximate generalized eigenfunctions of the continuous problem (Theorem 4.15). The next theorem shows that they even approximate eigenfunctions of the continuous problem.

Theorem 4.19 *Let $\mu(h)$ be an eigenvalue of T_h such that $\lim_{h\to 0} \mu(h) = \mu$. Suppose that for h small enough, w_h is a unit vector satisfying $(\mu(h) - T_h)^k w_h = 0$, where k is a positive integer with $k \leq \alpha_\mu$. Then, for any integer l with $k \leq l \leq \alpha_\mu$, there is a vector $u_{w_h} \in R(E)$ satisfying $(\mu - T)^l u_{w_h} = 0$ and*

$$\|u_{w_h} - w_h\| \leq C \|(T - T_h)|_{R(E)}\|^{(l-k+1)/\alpha_\mu},$$

where C is a constant C independent of h and l.

For the proofs of the last three theorems, we again refer to [75].

4.1.3 A Priori Error Estimates for the Finite Element Approximation

Approximation of Spectra

We start by deriving estimates of $\|T - T_h\|$ which allow us to apply the results of Section 4.1.2. In the following, we assume $\mathcal{T}_h$ to be a quasi uniform triangulation $\mathcal{T}_h$ of Ω, see [19]. Furthermore, let the finite element space $V_h \subset V \subset H^m(\Omega)$ be defined such that the restriction of any $v_h \in V_h$ on each cell $K \in \mathcal{T}_h$ is a polynomial of degree at most equal to $r - 1$. Then the following lemma holds, see [19].

Lemma 4.20 (Bramble-Hilbert) *Let $t \geq 2$. Then, under the assumptions made above, there exists a constant C such that*

$$\inf_{\chi \in V_h} \|v - \chi\|_j \leq Ch^{t-j} \|v\|_t \quad \text{for any } v \in H^t(\Omega), \quad 0 \leq j \leq t \leq r.$$

Furthermore, we assume that the operator A (see Section 4.1.1) is H^s-*regular*, i.e. for any $f \in H^{s-2m}$ and $u = A^{-1}f$, we have that $u \in H^s$, and there is a constant $C_s = C(\Omega, s)$ such that

$$\|u\|_s \leq C_s \|f\|_{s-2m}, \tag{4.13}$$

where $2m$ denotes the order of the elliptic operator $\mathcal{A}$ (see (4.1)). In our context, inequality (4.13) is assumed to hold for $s \leq r$. If we have sufficiently smooth coefficients $a_{\alpha\beta}$ and a sufficiently smooth boundary $\partial\Omega$ as well, one can prove the H^s-regularity of A under certain boundary conditions, see e.g. [50]. Note that H^s-regularity of A implies in particular that all eigenfunctions have H^s-regularity, see [50, 53].

Theorem 4.21 *Given $f \in H^t(\Omega)$ and $\varphi \in H^s(\Omega)$ with $0 \leq s, t \leq r - 2m$ where $r - 1$ is the polynomial order of the considered finite element discretization, we have*

$$|((T - T_h)f, \varphi)_0| \leq Ch^{t+s+2m} \|f\|_t \|\varphi\|_s.$$

4.1. Galerkin Finite Element Spectral Approximation

Proof. For any $\chi \in V_h$ we have

$$\begin{aligned}
\|(T-T_h)f\|_m^2 &\leq \frac{1}{C_E} |a((T-T_h)f,(T-T_h)f)| & \text{(ellipticity)} \\
&= |a((T-T_h)f, Tf-\chi)| & \text{(Galerkin orthogonality)} \\
&\leq \frac{C_B}{C_E} \|(T-T_h)f\|_m \|Tf-\chi\|_m & \text{(continuity).}
\end{aligned}$$

Hence,

$$\begin{aligned}
\|(T-T_h)f\|_m &\leq \frac{C_B}{C_E} \inf_{\chi \in V_h} \|Tf-\chi\|_m \\
&\leq \tilde{C} h^{t+2m-m} \|Tf\|_{t+2m} & \text{(Bramble-Hilbert)} \\
&\leq \hat{C} h^{t+m} \|f\|_t & \text{(regularity).} \quad (4.14)
\end{aligned}$$

With a *duality argument* (also known as the *Nitsche trick*) we derive an estimate in $L^2(\Omega)$: For any $\chi \in V_h$ we have

$$\begin{aligned}
|((T-T_h)f, \varphi)_0| &= |a((T-T_h)f, T^*\varphi)| \\
&= |a((T-T_h)f, T^* - \chi)| & \text{(Galerkin orthogonality)} \\
&\leq C_E \|(T-T_h)f\|_m \inf_{\chi \in V_h} \|T^*\varphi - \chi\|_m & \text{(continuity)} \\
&\leq C_E \hat{C} h^{t+m} \|f\|_t \inf_{\chi \in V_h} \|T^*\varphi - \chi\|_m & \text{(see (4.14))} \\
&\leq \overline{C} h^{t+m} \|f\|_t h^{s+m} \|T^*\varphi\|_{s+2m} & \text{(Bramble-Hilbert)} \\
&\leq C h^{t+s+2m} \|f\|_t \|\varphi\|_s & \text{(regularity),}
\end{aligned}$$

which completes the proof. □

Theorem 4.21 now allows us to prove assumption (4.12) we made in Section 4.1.2:

Corollary 4.22 T_h *converges to* T, *i.e.* $\|T-T_h\|_0 \to 0$ *as* $h \to 0$.

Proof. Let $g \in L^2(\Omega)$ and choose $t = s = 0$ in Theorem 4.21. Then we have

$$\|(T-T_h)g\|_0 = \sup_{\varphi \in L^2, \|\varphi\|_0 = 1} |((T-T_h)g, \varphi)_0| \leq C h^{2m} \|g\|_0.$$

□

Theorem 4.23 *Let* $r-1$ *be the degree of the polynomials considered in the finite element* *discretization.* Then the following estimates

$$\|(T-T_h)|_{R(E)}\|_0 \leq Ch^r \quad \text{and} \quad \|(T^*-T_h^*)|_{R(E)}\|_0 \leq Ch^r$$

hold.

Proof. Let $f \in R(E)$. The regularity assumption (4.13) implies $R(E) \subset H^{r-2m}(\Omega)$. Applying Theorem 4.21 with $t = r - 2m$ and $s = 0$, we have for any $\varphi \in L^2(\Omega)$

$$|((T-T_h)f, \varphi)_0| \leq \tilde{C}h^r \|f\|_{r-2m} \|\varphi\|_0,$$

hence,

$$\begin{aligned}
\|(T-T_h)|_{R(E)}\|_0 &= \sup_{f \in R(E), \|f\|_0 = 1} \|(T-T_h)f\|_0 \\
&= \sup_{f \in R(E), \|f\|_0 = 1} \sup_{\varphi \in L^2(\Omega), \|\varphi\|_0 = 1} |((T-T_h)f, \varphi)_0| \\
&\leq \tilde{C}h^r \sup_{f \in R(E), \|f\|_0 = 1} \|f\|_{r-2m} \\
&\leq Ch^r,
\end{aligned}$$

where the last estimate is derived by using that all norms are equivalent in the finite-dimensional spaces $R(E)$.

The second assertion can be proven analogously. $\square$

Now we are able to prove the main theorem which states quantitative convergence results for the eigenvalues.

Theorem 4.24 *Let $\mathcal{A}$ be a H^r-regular, uniformly elliptic operator of order $2m$ as defined in (4.1), where $r - 1$ denotes the order of the finite element approximation. Furthermore, assume λ to be an eigenvalue of $\mathcal{A}$ with algebraic multiplicity σ and ascent α. Then, for sufficiently small h, there are exactly σ approximating eigenvalues $\{\lambda_{h,i}\}_{i=1,\ldots,\sigma}$ of the discrete problem (4.9) counted according to their algebraic multiplicity such that*

$$\left|\lambda - \frac{1}{\sigma}\sum_{i=1}^{\sigma} \lambda_{h,i}\right| \leq C_\lambda h^{2(r-m)}, \tag{4.15}$$

$$|\lambda - \lambda_{h,i}| \leq C'_\lambda h^{2(r-m)/\alpha} \quad \text{for } i = 1, \ldots, \sigma, \tag{4.16}$$

where C_λ and C'_λ are constants independent of h.

4.1. Galerkin Finite Element Spectral Approximation

Proof. Theorem 4.21 with $t = s = r - 2m$ yields

$$\sum_{j=1}^{\sigma} |((T - T_h)\varphi_j, \varphi_j^*)_0| \leq C h^{2r-2m}. \tag{4.17}$$

The first assertion (4.15) follows from Theorem 4.17 combined with the estimate (4.17) and Theorem 4.23. The second assertion (4.16) follows analogously from Theorem 4.18 together with the estimate (4.17) and Theorem 4.23. □

By combining Theorem 4.15 and Theorem 4.23, we directly gain the following result about the convergence order of the gap between the generalized eigenspaces and their approximations.

Theorem 4.25 *For a finite element approximation of order $r - 1$ there holds for sufficiently small h*

$$\hat{\delta}_{L^2(\Omega)}(R(E), R(E_h)) \leq C h^r,$$

where C is a constant independent of h.

Finally, we state the qualitative convergence of the generalized eigenfunctions.

Theorem 4.26 *Let the assumptions of Theorem 4.24 hold. Furthermore, let A_h denote the linear bounded operator associated to the sesquilinear form in (4.7) and λ_h an eigenvalue of A_h converging to an eigenvalue λ of A (i.e. of (4.3)) with ascent α. Suppose that for h small enough, w_{u_h} is a unit vector such that $(\lambda_h - A_h)^k w_{u_h} = 0$ holds for any integer k with $1 \leq k \leq \alpha$. Then, for any integer l with $k \leq l \leq \alpha$ there exists a vector u_h satisfying $(\lambda - T)^l u_h = 0$ and*

$$\|u_h - w_{u_h}\|_0 \leq C h^{r(l-k+1)/\alpha},$$

where C is a constant independent of h and l.

Proof. The assertion follows directly from Theorem 4.19 and Theorem 4.23. □

Convergence results can also be stated in the H^m-norm by considering $\tilde{T}$ and its discrete counterpart $\tilde{T}_h$ for $V = \tilde{H}^m(\Omega)$ (cf. Definitions 4.1 and 4.3). We show this exemplarily for the convergence order of the gap between $R(E)$ and $R(E_h)$. Using inequality (4.14) we have that

$$\|(T - T_h)f\|_m \leq \hat{C} h^{t+m} \|f\|_t. \tag{4.18}$$

For $t = 0$ we obtain $\|(T - T_h)f\|_m \leq C h^m \|f\|_0 \leq C h^m \|f\|_0$ and consequently

$$\lim_{h \to 0} \|\tilde{T} - \tilde{T}_h\|_m = 0.$$

52 DISCRETIZATION OF THE EIGENVALUE PROBLEMS

Setting $t = r - 2m$ in (4.18) and using analogous arguments as in the proof of Theorem 4.23, we obtain
$$\|(T - T_h)|_{R(E)}\|_0 \leq Ch^{r-m}.$$
Thus, using Theorem 4.15 we conclude the corresponding formulation of Theorem 4.25:
$$\hat{\delta}_{H^m(\Omega)}\left(R(E), R(E_h)\right) \leq Ch^{r-m},$$
where E (and E_h) is the spectral projector associated to $\tilde{T}$ (and $\tilde{T}_h$ respectively).

Approximation of Pseudospectra

The previously stated error estimates for eigenvalues allow us to derive a priori error estimate for pseudospectra with respect to the spectral norm. We start with a theorem quantifying the spectral norm for compact operators. In this view, recall that by virtue of the Riesz-Schauder theory any compact operator possesses an eigenvalue with maximum absolute value, cf. Theorem 4.6.

Theorem 4.27 *Let $T : H \to H$ be a compact operator on a real or complex Hilbert space H. Then for the spectral norm $\|\cdot\|_0$ we have*
$$\|T\|_0 = s_{\max}(T),$$
where $s_{\max}(T)$ denotes the largest singular value of T, i.e. the squareroot of the largest eigenvalue of T^*T.

Proof. We adapt the proof for the finite-dimensional case which can be found e.g. in [98]. We have
$$\|T\|_0^2 = \sup_{\|x\|_0=1} (Tx, Tx)_0 = \sup_{\|x\|_0=1} (T^*Tx, x)_0,$$
where T^*T is a normal compact operator with real nonnegative eigenvalues. By the spectral theorem for compact operators (see e.g. [113]), there exists an orthonormal system $\{e_1, e_2, \ldots\}$ of eigenvectors associated to the non-zero eigenvalues $\{\lambda_1, \lambda_2 \ldots\}$ of T^*T such that
$$T^*Tx = \sum_k \lambda_k (x, e_k)_0 \, e_k$$
holds for any $x \in H$. Consequently,
$$\|T\|_0^2 = \sup_{\|x\|_0=1} \sum_k \lambda_k (x, e_k)_0 (e_k, x)_0 = \sup_{\|x\|_0=1} \sum_k \lambda_k \, |(e_k, x)|^2$$
$$\leq \lambda_{\max}(T^*T) \sup_{\|x\|_0=1} \sum_k |(e_k, x)|^2,$$

4.1. Galerkin Finite Element Spectral Approximation

and applying Parseval's identity we conclude

$$\|T\|_0 \leq s_{\max}(T).$$

On the other hand, we derive

$$\|T\|_0^2 = \sup_{\|x\|_0=1} (T^*Tx, x)_0 \geq (T^*Te_{\max}, e_{\max})_0 = \lambda_{\max}(T^*T),$$

where $e_{\max}$ denotes an eigenvector associated to the largest eigenvalue $\lambda_{\max}(T^*T)$. This completes the proof. $\square$

Theorem 4.27 implies

$$\|A^{-1}\|_0 = \frac{1}{s_{\min}(A)},$$

for a bounded operator A with a compact inverse.

For any $z \in \mathbb{C}$, shifting the elliptic operator $\mathcal{A}$ in (4.1) to $\mathcal{A} - z\mathcal{I}$ clearly leaves the operator elliptic. For $\mu \notin \sigma(A) \cap \sigma(A_h)$ we can define the shifted version T^μ (and T_h^μ) of the compact operator T (and T_h) analogously to Definition 4.1 (and Definition 4.3).

Definition 4.28 *For $\mu \notin \sigma(A) \cap \sigma(A_h)$ we define*

$$T^\mu : L^2(\Omega) \to L^2(\Omega), \qquad T = I_{V \hookrightarrow L^2(\Omega)} \circ (A - \mu I_{V \hookrightarrow V'})^{-1} \circ I_{L^2(\Omega) \hookrightarrow V'},$$

$$T_h^\mu : L^2(\Omega) \to L^2(\Omega), \qquad T = I_{V_h \hookrightarrow L^2(\Omega)} \circ (A_h - \mu I_{V_h \hookrightarrow V'})^{-1} \circ I_{L^2(\Omega) \hookrightarrow V'}$$

Now we are able to state the main theorem for the convergence of the resolvent norm.

Theorem 4.29 *Let $\mathcal{A}$ be a H^r-regular, uniformly elliptic operator of order $2m$ as defined in (4.1), where $r - 1$ denotes the order of the finite element approximation. Further, let $\mu \notin \sigma(A) \cap \sigma(A_h)$ and α denote the ascent of the largest singular value ς of T^μ, i.e. the ascent of the largest eigenvalue of $(T^\mu)^*T^\mu$. Then, for h small enough, we have*

$$|\|T^\mu\|_0 - \|T_h^\mu\|_0| \leq C_\varsigma h^{2(r-m)/\alpha}, \tag{4.19}$$

where C_ς is a constant independent of h.

Proof. This assertion follows by the proof of Theorem 4.24 applied on the compact operators $(T^\mu)^*T^\mu$ and $(T_h^\mu)^*T_h^\mu$ to approximate the largest eigenvalue of $(T^\mu)^*T^\mu$. $\square$

4.1.4 Bibliographical Remarks

There are many publications on the spectral approximation theory for elliptic operators. In addition to the references already cited we mention here [9, 20, 50, 76]. For a posteriori error estimates we refer to [53, 56, 57]. The case of noncompact operators is described in [34, 35]. Mixed and hybrid methods are treated in [68]. Without any attempt of completeness, we mention [2, 61] as further works on spectral approximation. For treatises on finite element methods we refer to [19, 21, 29, 47, 81].

4.2 Discretization of the Neutron Transport Equation

In the setup of general reactors no exact solutions of the transport equation are known. This is especially due to the vast amount of detailed information about neutron cross sections which has to be known. Thus, sophisticated approximation methods and numerical methods are necessary. There exist two general approaches: *deterministic* and *Monte Carlo* methods. Monte Carlo methods are stochastic methods that simulate a finite number of particle histories. In this work we confine ourselves to deterministic methods and refer to [13, 66] for a discussion of Monte Carlo methods with respect to neutron transport problems.

The criticality problem involves three independent variables, namely, the position $\mathbf{r}$, the direction of travel $\mathbf{\Omega}$, and the energy E. In the approach used in this work, the position dependence is approximated by means of finite element methods. The dependence on particle direction is expanded as a series of spherical harmonics, whereas the energy dependence is approximated by piecewise constant functions (*multigroup approximation*).

In this work we restrict ourselves to the k eigenvalue problem (3.4). The discretization of the time dependent transport equation and the α eigenvalue problem are treated elsewhere, see [66] and [63] respectively.

4.2.1 Energy Discretization

In order to discretize the energy variable, the energy interval of interest is divided into a finite number of intervals (or *groups*) (E_g, E_{g-1}), $g = 1, \ldots, G$. The cross section in each group is assumed to be independent of the energy. This may be achieved for instance by averaging within each interval. Note that for increasing g the energy of the associated group is decreasing.

We approximate the angular flux within every group g by a function $\psi_g(\mathbf{r}, \mathbf{\Omega})$ known

4.2. Discretization of the Neutron Transport Equation

as the *group angular flux*:

$$\psi_g(\mathbf{r}, \mathbf{\Omega}) = \int_g \psi(\mathbf{r}, \mathbf{\Omega}, E)\, dE,$$

where integration over g means integration over the interval (E_g, E_{g-1}).

We assume for now that the energy dependence is separable, i.e. that the angular flux can be written as a product of a known function $f(E)$ and the group angular flux:

$$\psi(\mathbf{r}, \mathbf{\Omega}, E) = f(E)\psi_g(\mathbf{r}, \mathbf{\Omega}). \tag{4.20}$$

Note that by definition of the group angular flux the function $f(E)$ is normalized in the sense that

$$\int_g f(E)\, dE = 1$$

holds for any group g.

Integrating (3.4) between E_g and E_{g-1}, and substituting (4.20) in the resulting equations yields

$$\left[\mathbf{\Omega} \cdot \nabla + \sigma^g(\mathbf{r})\right]\psi_g(\mathbf{r}, \mathbf{\Omega}) = \sum_{g'=1}^{G} \int_S \sigma_s^{g' \to g}(\mathbf{r}, \mathbf{\Omega}' \cdot \mathbf{\Omega})\, \psi_{g'}(\mathbf{r}, \mathbf{\Omega}')\, d\mathbf{\Omega}' + \frac{1}{k}\mathbf{F}\psi \tag{4.21}$$

for any $g = 1, \ldots, G$, where

$$\mathbf{F}\psi = \frac{\chi_g}{k} \sum_{g'=1}^{G} \nu_{g'}\, \sigma_f^{g'}(\mathbf{r}) \int_S \psi_{g'}(\mathbf{r}, \mathbf{\Omega}')\, d\mathbf{\Omega}',$$

and $\psi = (\psi_1, \psi_2, \ldots, \psi_G)^T$. The multigroup cross sections in (4.21) are defined by

$$\sigma^g(\mathbf{r}) = \int_g \sigma(\mathbf{r}, E) f(E)\, dE,$$

$$\sigma_s^{g' \to g}(\mathbf{r}, \mathbf{\Omega}' \cdot \mathbf{\Omega}) = \int_g \int_{g'} \sigma_s(\mathbf{r}, E' \to E, \mathbf{\Omega}' \cdot \mathbf{\Omega}) f(E')\, dE'\, dE,$$

$$\nu_{g'}\, \sigma_f^{g'}(\mathbf{r}) = \int_{g'} \nu(E) \sigma_f(\mathbf{r}, E) f(E)\, dE,$$

and moreover we have set

$$\chi_g = \int_g \chi(E)\, dE.$$

The separability assumption (4.20) is in fact not needed to derive the *within group* equations (4.21). By dropping this assumption one even obtains improved definitions of the multigroup cross sections. However, we skip this lengthy derivation and refer to

[13, 14, 66] for the interested reader.

We define the *streaming collision operator* for a group g as

$$H^0_{gg}\psi_g = [\mathbf{\Omega} \cdot \nabla + \sigma^g(\mathbf{r})]\,\psi_g(\mathbf{\Omega}, \mathbf{r})$$

and the *group-to-group scattering operator* as

$$H^1_{gg'}\psi_{g'} = \int_S \sigma_s^{g' \to g}(\mathbf{r}, \mathbf{\Omega}' \cdot \mathbf{\Omega})\,\psi_g(\mathbf{r}, \mathbf{\Omega}')\,d\mathbf{\Omega}'.$$

Moreover, the *multigroup transport operator* is defined by

$$H_{gg'} = \delta_{gg'} H^0_{gg} - H^1_{gg'}.$$

Setting the block matrix $\mathbf{H}$ as

$$\mathbf{H} = \begin{bmatrix} H_{11} & H_{12} & \cdots & H_{1G} \\ H_{21} & H_{22} & \cdots & H_{2G} \\ \vdots & & \ddots & \\ H_{G1} & & & H_{GG} \end{bmatrix},$$

we obtain the block wise representation

$$\mathbf{H}\psi = \frac{1}{k}\mathbf{F}\psi \qquad (4.22)$$

of the criticality problem.

4.2.2 Second Order Even Parity Formulation

The second order even parity formulation allows us to derive self-adjoint diagonal blocks H_{gg} in the block matrix $\mathbf{H}$. We start with the within-group equations (4.21), where we assume for simplification the scattering multigroup cross section to be isotropic. For anisotropic scattering the derivation becomes more cumbersome and can for instance be found in [65, 106]. Furthermore, for brevity we assume to have only one energy group. This means we can drop the group indices, and consequently equation (4.21) reads

$$\left[\mathbf{\Omega} \cdot \nabla + \sigma(\mathbf{r})\right]\psi(\mathbf{r}, \mathbf{\Omega}) = \sigma_s(\mathbf{r})\int_S \psi(\mathbf{r}, \mathbf{\Omega}')\,d\mathbf{\Omega}' + \frac{\chi}{k}\nu\,\sigma_f(\mathbf{r})\int_S \psi(\mathbf{r}, \mathbf{\Omega}')\,d\mathbf{\Omega}'. \qquad (4.23)$$

We proceed by splitting the angular flux into even and odd angular parity components

$$\psi(\mathbf{r}, \mathbf{\Omega}) = \psi^+(\mathbf{r}, \mathbf{\Omega}) + \psi^-(\mathbf{r}, \mathbf{\Omega}),$$

4.2. Discretization of the Neutron Transport Equation

where the even parity component ψ^+ and the odd parity component ψ^- are defined by

$$\psi^{\pm}(\mathbf{r},\mathbf{\Omega}) = \frac{1}{2}\left[\psi(\mathbf{r},\mathbf{\Omega}) \pm \psi(\mathbf{r},-\mathbf{\Omega})\right].$$

Then, we evaluate (4.23) at $-\mathbf{\Omega}$ to gain

$$\left[-\mathbf{\Omega}\cdot\nabla + \sigma(\mathbf{r})\right]\psi(\mathbf{r},-\mathbf{\Omega}) = \sigma_s(\mathbf{r})\int_S \psi(\mathbf{r},\mathbf{\Omega}')\,d\Omega' + \frac{\chi}{k}\nu\sigma_f(\mathbf{r})\int_S \psi(\mathbf{r},\mathbf{\Omega}')\,d\Omega'. \quad (4.24)$$

Adding equations (4.23) and (4.24) yields

$$\mathbf{\Omega}\cdot\nabla\psi^-(\mathbf{r},\mathbf{\Omega}) + \sigma(\mathbf{r})\psi^+(\mathbf{r},\mathbf{\Omega}) = \sigma_s(\mathbf{r})\int_S \psi(\mathbf{r},\mathbf{\Omega}')\,d\Omega' + \frac{\chi}{k}\nu\sigma_f(\mathbf{r})\int_S \psi(\mathbf{r},\mathbf{\Omega}')\,d\Omega'.$$

Analogously, by subtracting these two equations we have

$$\mathbf{\Omega}\cdot\nabla\psi^+(\mathbf{r},\mathbf{\Omega}) + \sigma(\mathbf{r})\psi^-(\mathbf{r},\mathbf{\Omega}) = 0.$$

Finally, we combine the two last equations to eliminate ψ^- which results in the second order even parity formulation of the criticality problem:

$$-\mathbf{\Omega}\cdot\nabla\left[\frac{1}{\sigma(\mathbf{r})}\mathbf{\Omega}\cdot\nabla\psi^+(\mathbf{r},\mathbf{\Omega})\right] + \sigma(\mathbf{r})\psi^+(\mathbf{r},\mathbf{\Omega}) = \sigma_s(\mathbf{r})\int_S \psi^+(\mathbf{r},\mathbf{\Omega}')\,d\Omega' \\ + \frac{\chi}{k}\nu\sigma_f(\mathbf{r})\int_S \psi^+(\mathbf{r},\mathbf{\Omega}')\,d\Omega'. \quad (4.25)$$

Note that since $\psi^+(\mathbf{r},\mathbf{\Omega}) = \psi^+(\mathbf{r},-\mathbf{\Omega})$, the even parity formulation (4.25) only needs to be solved over half of the angular domain.

The vacuum boundary condition for the angular flux can be shown to result in a boundary condition for the even parity component:

$$\mathbf{\Omega}\cdot\nabla\psi^+(\mathbf{r},\mathbf{\Omega}) \pm \sigma(\mathbf{r})\psi^+(\mathbf{r},\mathbf{\Omega}) = 0, \quad \mathbf{n}\cdot\mathbf{\Omega} \gtrless 0, \quad \mathbf{r}\in\Gamma,$$

see [66]. The reflecting boundary conditions for the even parity component are given by

$$\psi^+(\mathbf{r},\mathbf{\Omega}) = \psi^+(\mathbf{r},\mathbf{\Omega}'), \quad \mathbf{n}\cdot\mathbf{\Omega} = -\mathbf{n}\cdot\mathbf{\Omega}', \quad (\mathbf{\Omega}\times\mathbf{\Omega}')\cdot\mathbf{n} = 0, \quad \mathbf{r}\in\Gamma,$$

where $\mathbf{\Omega}'$ denotes the incident direction and $\mathbf{\Omega}$ the reflection direction.

4.2.3 Angular and Spatial Discretization

The angular variable is treated mainly using three types of techniques. *Integral methods* are based on integrating out the angular dependence from the transport equation, but they lead to dense matrices. The *Discrete ordinate* (or S_N) *collocation method* consists in evaluating the neutron transport equation along a finite number of angular directions. This methods is simple to use but it suffers from anomalies in the angular flux distribution known as *ray effects*, see [66].

In this work we consider the *spherical harmonic* (or P_N) *method* which does not suffer from the ray effect phenomenon. The spherical harmonics are based on Legendre polynomials and form an orthogonal basis of the square integrable functions on the unit sphere. The idea relies on expanding the angular dependence in a series of spherical harmonics truncated at order N.

The normalized spherical harmonics are defined by

$$Y_{lm}(\mathbf{\Omega}) = Y_{lm}(\vartheta, \varphi) = \sqrt{\frac{2l+1}{4\pi} \frac{(l-m)!}{(l+m)!}} P_{lm}(\cos\vartheta) e^{im\varphi}, \quad l = 0, 1, \ldots, \quad m = 0, 1, \ldots, l,$$

where P_{lm} are the associated Legendre polynomials, $\vartheta \in [0, \pi]$ denotes the polar angle, and $\varphi \in [0, 2\pi]$ the azimuthal angle. These functions are orthonormal in the sense that

$$\int_S \overline{Y_{lm}(\mathbf{\Omega})} Y_{l'm'}(\mathbf{\Omega}) \, d\mathbf{\Omega} = \int_0^{2\pi} \int_0^{\pi} \overline{Y_{lm}(\vartheta,\varphi)} Y_{l'm'}(\vartheta,\varphi) \sin\vartheta \, d\vartheta \, d\varphi = \delta_{ll'} \delta_{mm'}.$$

Moreover, they satisfy the parity property

$$Y_{lm}(-\mathbf{\Omega}) = Y_{lm}(\pi - \vartheta, \pi + \varphi) = (-1)^l Y_{lm}(\vartheta,\varphi) = (-1)^l Y_{lm}(\mathbf{\Omega}).$$

Hence, for expanding the even parity angular flux, only components Y_{lm} with l even are needed. Splitting Y_{lm} into real and imaginary parts yields

$$Y_{lm}(\mathbf{\Omega}) = Y_{lm}^e(\mathbf{\Omega}) + i Y_{lm}^o(\mathbf{\Omega}),$$

where Y_{lm}^e is even in φ and Y_{lm}^o odd in φ.

The P_N method for even parity in $\mathbf{\Omega}$ uses the real and the imaginary part of Y_{lm} as basis functions, where l is even with $l < N$. We define the vector

$$\mathbf{y}(\mathbf{\Omega}) = \left(Y_{00}, Y_{20}, Y_{21}^e, Y_{21}^o, Y_{22}^e, Y_{22}^o, Y_{40}, \ldots, Y_{N-1\,N-1}^e, Y_{N-1\,N-1}^o \right)^T$$

containing the basis functions for the angular dependence.

4.2. Discretization of the Neutron Transport Equation

The dependence on the spatial variable is treated by means of finite element methods. Let $\mathbf{f}$ denote the vector containing a finite element basis of dimension s, i.e.

$$\mathbf{f}(\mathbf{r}) = (f_1(\mathbf{r}), f_2(\mathbf{r}), \ldots f_s(\mathbf{r}))^T.$$

Then, the spatio-angular discretization ψ_h^+ of ψ^+ is given by

$$\psi_h^+(\mathbf{r}, \mathbf{\Omega}) = \sum_{i,j} \alpha_{ij} f_i(\mathbf{r}) y_j(\mathbf{\Omega}), \qquad (4.26)$$

where the coefficients α_{ij} are to be determined. Here, $y_j(\mathbf{\Omega})$ denotes the j-th component of the vector $\mathbf{y}(\mathbf{\Omega})$.

In order to derive the weak matrix formulation of the criticality problem, one plugs the spatio-angular discretization (4.26) in the even parity formulation of (4.21), where most commonly the scattering cross section is expanded in Legendre polynomials. Then, the resulting equation is tested by the expansion functions of ψ_h^+, i.e. we multiply by functions which are of the form $f_i(\mathbf{r}) y_j(\mathbf{\Omega})$ and integrate over space and angle. This results in a matrix form of a generalized eigenvalue problem seeking the largest eigenvalue λ (which corresponds to the k eigenvalue to be approximated) and the associated eigenvector u (which holds the coefficients α_{ij} of ψ_h^+ for each energy group) of

$$Fu = \lambda H u.$$

Though straightforward, we do not present these tedious calculations and refer instead to [106] for a detailed description of the weak form and its matrix formulation.

Chapter 5

Eigenvalue Solvers and Parallelization

The eigenvalue problems arising in the fields of hydrodynamic stability and criticality are highly complex. This fact is especially reflected by the large size and complexity of the resulting discretized problems. Therefore, we need powerful eigenvalue solvers which in addition allow an efficient parallelization. In this work we focus on the application of the Davidson method combined with different preconditioning techniques to compute pseudospectra of elliptic operators and k eigenvalues of the linear Boltzmann equation.

In the following section we establish the Davidson method with a brief review of eigenvalue projection methods. Thereafter, in Section 5.2, we focus on pseudospectra computation in the spectral norm. In Section 5.3 we treat the parallelization schemes of basic linear algebra routines in terms of finite element discretizations as well as an additional parallelization technique for the computation of pseudospectra.

5.1 The Davidson Method

The *Davidson method* is an iterative algorithm to compute few extrem eigenvalues and the corresponding eigenvectors of large sparse matrices. It was originally developed by Davidson [32] for real-symmetric matrices in the field of quantum chemistry. Later, this method was also successfully applied to nonsymmetric problems, see [69, 88]. It is closely related to Krylov subspace methods which are among the most important methods for solving large sparse eigenvalue problems. The concept of Krylov methods relies on the projection of the problem onto a smaller subspace.

Let $A \in \mathbb{C}^{n \times n}$ and $\mathcal{K}, \mathcal{L}$ be two subspaces of $\mathbb{C}^n$ of same dimension m. We seek for $\lambda \in \mathbb{C}$ and $u \in \mathbb{C}^n \backslash \{0\}$ such that

$$Au = \lambda u. \qquad (5.1)$$

The concept of projection methods is to approximate the exact eigenvector u by a vector

Algorithm 5.1 Power method.

1: **function** POWER(A, k)
2:     Choose an initial normalized vector v_0;
3:     **for** $k = 1, 2, \ldots$ **do**
4:         Compute $z_k = Av_{k-1}$;
5:         Compute $v_k = z_k/\|z_k\|_2$;
6:         Compute the Rayleigh quotient $\alpha_k = v_k^H A v_k$;
7:     **end for**
8: **end function**

$\tilde{u}$ belonging to the subspace $\mathcal{K}$ and the corresponding eigenvalue λ by some $\tilde{\lambda}$ such that the *Petrov-Galerkin* condition

$$A\tilde{u} - \tilde{\lambda}\tilde{u} \perp \mathcal{L} \tag{5.2}$$

is satisfied. Let $V = [v_1, v_2, \ldots, v_m]$ denote the matrix whose column vectors v_i form a basis of $\mathcal{K}$ and $W = [w_1, w_2, \ldots, w_m]$ the matrix whose column vectors w_i form a basis of $\mathcal{L}$. Suppose that V and W are biorthogonal, i.e. $W^H V = I$, where I is the m-dimensional identity matrix. Then, we can reformulate the projection method by seeking $\tilde{\lambda} \in \mathbb{C}$ and $y \in \mathbb{C}^m \setminus \{0\}$ satisfying

$$H_m y = \tilde{\lambda} y, \tag{5.3}$$

where $H_m = W^H A V \in \mathbb{C}^{m \times m}$ is the *Rayleigh matrix*. Since we have usually $m \ll n$, solving the eigenvalue problem (5.3) is much less expensive compared to solving the original problem (5.1).

In the case $\mathcal{L} \neq \mathcal{K}$, projection methods are referred to as *oblique projection methods*, whereas in the case $\mathcal{L} = \mathcal{K}$, they are referred to as *orthogonal projection methods*. General convergence results of projection methods can be found in [86].

A very simple projection method is obtained by choosing

$$\mathcal{L} = \mathcal{K} = \operatorname{span}\{A^k v\}$$

for an integer $k \geq 0$. The resulting procedure is known as *power method* and given in Algorithm 5.1, see e.g. [49]. At each iteration k, the eigenvalue problem is projected onto the one-dimensional subspace spanned by $A^k v_0$. Under very mild conditions the sequence α_k converges to the largest eigenvalue in modulus λ_1 of A and the sequence v_k to the corresponding eigenvector. This holds, if λ_1 is dominant (i.e. there is one and only one eigenvalue of largest modulus) and semi-simple, and if v_0 has a component of the eigenvector associated with this eigenvalue, see [86]. The convergence rate is dictated by the quotient $|\lambda_2|/|\lambda_1|$, where λ_2 is the second largest eigenvalue in modulus.

More sophisticated projection methods can be derived by employing *Krylov subspaces*

5.1. The Davidson Method

which are defined by

$$\mathcal{K}_m = \mathcal{K}_m(A, v) = \text{span}\{v, Av, A^2v, \ldots, A^{m-1}v\}$$

for a given $v \in \mathbb{C}^n$. Krylov subspaces have desirable properties in the context of projection methods, see e.g. [86, 87]. For instance, if we choose both $\mathcal{L}$ and $\mathcal{K}$ to be the Krylov subspace $\mathcal{K}_m(A, v)$, the projected matrix H_m is an upper Hessenberg matrix, i.e. it has zero entries below the first subdiagonal. In this case, the resulting projection method is known as *Arnoldi method*. If we further assume A to be Hermitian, i.e. $A^H = A$, the Rayleigh matrices H_m are tridiagonal. Then the Arnoldi algorithm simplifies to the so-called *Lanczos method*.

To speed up convergence one may employ preconditioning techniques. The convergence properties of the Arnoldi method are directly linked to the separation properties of the eigenvalues, see [86]. A better separation of the desired eigenvalue leads to a faster convergence. Therefore, one may reformulate (5.1) to an eigenvalue problem with better separation properties. For instance, one may transform (5.1) to $(A - \sigma I)^{-1}x = \mu x$, where μ is close to the desired eigenvalue λ. The eigenvectors of A and $(A - \sigma I)^{-1}$ are identical and the eigenvalues λ of A can be recovered by the identity $\mu = (\lambda - \sigma)^{-1}$. This is the concept of the *shift-and-invert Arnoldi method*.

A more general approach is chosen in the method of Davidson. It is based on a preconditioned version of the Arnoldi method, but in order to gain more flexibility it allows the preconditioning to vary at each iteration. Note that in this case the Rayleigh matrices H_m may become dense.

A block version of the Davidson method to compute l desired (for instance the smallest) eigenvalues λ_i of a matrix A is given in Algorithm 5.2, see [30, 86]. The integer m refers to the restart parameter and determines the maximum size of the basis V_j from which the eigenvectors are built. At each iteration k, we compute the Rayleigh projection H_k of the matrix A. Then, in step 5, the projected eigenvalue problem is solved. This problem is of maximum size $m + l$ and one can apply for instance the QR algorithm to determine the desired eigenvalues and the corresponding eigenvectors, see e.g. [49]. In step 11 (and step 13 respectively) new directions $t_{j,i} = C_{j,i}r_{j,i}$ are incorporated in the basis V_j. We choose the preconditioning matrices $C_{j,i}$ to be of the form M_i^{-1}, where M_i is an approximation of $(A - \lambda_i I)$. In this context, the abbreviation MGS stands for the modified Gram-Schmidt orthogonalization procedure.

Notice that besides some basic vector routines, only sparse matrix-vector multiplications Av for a given vector v are needed. Since the preconditioning matrices $C_{k,i}$ may vary at each iteration k, these need not to be built actually. Instead, one may use iterative methods in order to determine $t_{k,i}$, i.e. one solves $M_i t_{k,i} = r_{k,i}$ in step 9 iteratively.

Algorithm 5.2 Block version of Davidson method.

1: **function** DAVIDSON(A, m, l)
2:     Choose an initial orthonormal matrix $V_1 = [v_1, \ldots, v_l]$;
3:     **for** $k = 1, 2, \ldots$ **do**
4:         Compute the Rayleigh matrix $H_k = V_k^H A V_k$;
5:         Compute the l desired eigenpairs $(\tilde{\lambda}_{k,i}, y_{k,i})_{1 \leq i \leq l}$ of H_k;
6:         Compute the Ritz vectors $\tilde{u}_{k,i} = V_k y_{k,i}$ for $i = 1, \ldots, l$;
7:         Compute the residuals $r_{k,i} = A\tilde{u}_{k,i} - \tilde{\lambda}_{k,i}\tilde{u}_{k,i}$ for $i = 1, \ldots, l$;
8:         **if** convergence **then** exit;
9:         Compute new directions $t_{k,i} = C_{k,i} r_{k,i}$ for $i = 1, \ldots, l$;
10:        **if** $\dim(V_k) \leq m$ **then**
11:            $V_{k+1} = \text{MGS}(V_k, t_{k,1}, \ldots, t_{k,l})$;
12:        **else**
13:            $V_{k+1} = \text{MGS}(\tilde{u}_{k,1}, \ldots, \tilde{u}_{k,l}, t_{k,1}, \ldots, t_{k,l})$;
14:        **end if**
15:     **end for**
16: **end function**

Furthermore, note that the Rayleigh matrices H_k can be computed only by updating the last l rows and columns at each iteration k.

Algorithm 5.2 is valid for both symmetric and unsymmetric eigenvalue problems. The only difference is, that in the unsymmetric case, complex eigenvalues may occur. Nevertheless, a complex arithmetic is easily avoided by splitting complex vectors into two real vectors: one holding the real part, the other the imaginary part. In order to incorporate a complex vector to the basis V_j, we just append both real and imaginary part as two real vectors. This separation is no restriction. Rather, it induces more flexibility, since with more vectors in V_j, there are more ways to combine them.

The Davidson method can easily be extended to solve generalized eigenvalue problems

$$Au = \lambda M u.$$

If M is symmetric positive definite, one may use an M-orthogonalization procedure in step 11 and step 13 in Algorithm 5.2. For general M, we project both matrices A and M and solve a generalized eigenvalue problem in step 5 (for instance with the QZ algorithm, see [49]). In both cases, the residual computation in step 7 is replaced by $r_{k,i} = A\tilde{u}_{k,i} - M\tilde{\lambda}_{k,i}\tilde{u}_{k,i}$.

Convergence results for the symmetric case can be found [30, 70, 86]. The nonsymmetric case is examined in [69, 88]. An advancement of the Davidson method is the so-called *Jacobi-Davidson method*. It adds an additional oblique projection in step 7 of Algorithm 5.2, see [92]. However, the Jacobi-Davidson method is not proven to be better than the Davidson method in all cases as pointed out in [72].

5.2 Computation of Pseudospectra

In this work we consider pseudospectra with respect to the spectral norm. Then, pseudospectra can be determined by means of smallest singular values of shifted matrices, cf. Section 2.3.1. More precisely, we compute the norm of the resolvent $\|(z - A)^{-1}\|_2 = s_{\min}(z - A)$ for different z on a grid in the complex plane.

For large matrices, a complete singular value decomposition induces high computational cost and high storage requirements. Therefore, different computational methods have been developed, see [102, 103] and references therein.

The concept of the *poor man's pseudospectrum* relies on the equivalent definition based on disturbing the matrix A

$$\sigma_\varepsilon(A) = \bigcup_{\|E\| < \varepsilon} \sigma(A + E),$$

see Definition 2.2. The resulting algorithm computes spectra of the disturbed matrices $A + E$ for some random matrices E with $\|E\| < \varepsilon$. As we have seen in the proof of Theorem 2.3, it is sufficient that E is of rank 1 which reduces the cost of scaling this matrix, see [84].

Especially for smaller matrices the computation of pseudospectra can be performed by means of spectral dichotomy methods, see [48, 64]. First, ε-spectral spots, i.e. regions in the complex plain which contain the ε-pseudospectrum, are determined. Then, spectral projectors associated to these regions are computed. In addition to the ε-pseudospectrum, this method provides spectral projectors and invariant subspaces as well. However, it entails higher computational costs and is therefore preferred for small-scaled problems.

Another approach, in particular for large matrices, is to project the matrix A onto a subspace and compute the desired pseudospectra of the projected matrix. Let U denote a matrix whose columns form an orthonormal basis of the subspace under consideration. If the subspace is chosen to be invariant, it is easy to show that

$$\sigma_\varepsilon(U^H A U) \subseteq \sigma(A), \tag{5.4}$$

see [103]. In [101] a projection onto Krylov subspaces is proposed. Typically, the eigenvalues of the projected matrix $U^H A U$ converge to the eigenvalues of A close to the boundary of the spectrum. However, Krylov subspaces are not invariant in general and hence inclusion (5.4) does not hold. Nevertheless, by appending one row to the Hessenberg matrix $U^H A U$ one may show that the resulting pseudospectra are contained in $\sigma_\varepsilon(A)$, see [101]. In this case, the ε-pseudospectrum of a rectangular matrix is defined as the ε-curve of $\|(z - A)^{-1}\|$ with $(z - A)^{-1}$ denoting the pseudoinverse of $(z - A)$.

In this work we focus on the computation of pseudospectra by means of iterative sparse

Algorithm 5.3 Computation of a spectral portrait.
1: **function** DRAW_PORTRAIT$(A, (x_1, x_2), (y_1, y_2), nx, ny)$
2:     $h_x = \frac{x_2 - x_1}{nx - 1}, \quad h_y = \frac{y_2 - y_1}{ny - 1};$
3:     $z = x_1 + iy_1;$
4:     **for** $j = 1, \ldots, nx$ **do**
5:         **for** $k = 1, \ldots, ny$ **do**
6:             Compute $s_{\min}(z - A)$ with Algorithm 5.2;
                // *Start with the previous computed singular subspace*
7:             $z = z + ih_y;$ // *Next line*
8:         **end for**
9:         $z = z - ih_y + h_x;$ // *Next column*
10:        $h_y = -h_y;$ // *Change sweep direction along the column*
11:    **end for**
12: **end function**

eigenvalue solvers. In this context the Davidson method was successfully used, see [55, 80].

Following the definition of pseudospectra introduced in Section 2.3.1, we define the *spectral portrait* of a matrix A by the plot of the map

$$z \mapsto sp_{(A)}(z) = \log_{10}\left[\|(z - A)^{-1}\|_2\right] = -\log_{10}\left[s_{\min}(z - A)\right],$$

where $s_{\min}(z - A)$ denotes the smallest singular value of $z - A$.

The computation of a spectral portrait of a matrix A is outlined in Algorithm 5.3. Please note that here i refers to the imaginary unit. First, we define a grid in the domain of interest in the complex plane. The grids we consider are given in a rectangular domain $[x_1, x_2] \times [y_1, y_2]$ with $nx \times ny$ points (nx in the horizontal, ny in the vertical). Then, for any grid point z we compute $s_{min}(z - A) = \left[\lambda_{min}\left((z - A)^H(z - A)\right)\right]^{1/2}$ by means of the Davidson method. For two neighboring grid points z_1 and z_2, we expect the matrices $(z_1 - A)$ and $(z_2 - A)$ to have close singular values and close singular vectors. To improve performance, we start the Davidson algorithm using the singular subspace computed in the last step as initial guess, see line 6 in Algorithm 5.3. Note that we do not need to compute $(z - A)^H(z - A)$ explicitly. Only matrix vector products of the form Av and $A^H v$ for a given vector v are needed. Moreover, if the matrix A is real, the spectral portrait of A is clearly symmetric with respect to the real axis.

In hydrodynamic stability we are mainly interested in the spectral portrait around the origin. Since we expect the eigenvalues of A to be in this region, the smallest singular value of $z - A$ is likely close to zero. Therefore, we choose all the preconditioning matrices $C_{k,i}$ in the Davidson method to be of the form M^{-1}, where M is a approximation of $A^H A$. It is easy to choose M to be positive definite which is a desired property in the convergence statements of the Davidson method, see [23, 80].

For a matrix pencil (A, M), we define the spectral portrait as

$$z \mapsto gsp_{(A,M)}(z) = \log_{10}\left[\|(zM - A)^{-1}\|_2\right] = -\log_{10}\left[s_{\min}(zM - A)\right],$$

cf. Section 2.3.4. In this case, the computation of a spectral portrait is given by Algorithm 5.3, where we have only to replace line 6 by applying the Davidson method on $(zM - A)$.

5.3 Parallelization

The problems we have to cope with are usually of high order ($\sim 10^5 - 10^7$) and consequently entail high computational costs. To solve these kind of problems in reasonable time, parallelization techniques are virtually mandatory.

In this section we treat parallelization schemes on platforms with distributed memory. Since the most time consuming part in iterative solvers is the sparse matrix vector multiplication, we focus on this issue. We also show how this can be combined with a second level of parallelism in order to manage the very expensive computation of pseudospectra.

5.3.1 Sparse Matrix Vector Multiplication for Finite Element Methods

In our framework assembled matrices and vectors are distributed row-wise over the processes according to the distribution of the degrees of freedom (DoF) from the finite element approach. Each local sub-matrix is divided into two blocks: a diagonal block representing all couplings and interactions within the subdomain, and an off-diagonal block representing the couplings across subdomain interfaces.

In Figure 5.1 we find a domain partitioning into four subdomains. In order to determine the structure of the global system matrices, first, each subdomain has to be associated with a single process. Then, each process not only detects couplings within its own subdomain, but also couplings to the so-called *ghost DoF*, i.e. neighboring DoF which are owned by a different process. In terms of a global system matrix, DoF i and j interact if the matrix has a non-zero element in row i and column j.

The distributed sparse matrix vector multiplication is given in Algorithm 5.4. While every process is computing its local contribution of the matrix vector multiplication, an asynchronous communication for exchanging the ghost values is initiated. After this communication phase has been completed, the local contributions from the coupled ghost DoF are added accordingly, see [6, 58].

EIGENVALUE SOLVERS AND PARALLELIZATION

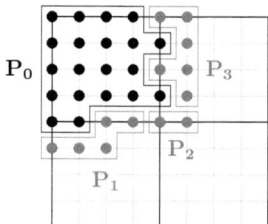

Figure 5.1: Domain partitioning: DoF of process $\mathbf{P_0}$ are marked in black (interior DoF in diagonal block); the remaining DoF marked in gray represent inter-process couplings for process $\mathbf{P_0}$ (ghost DoF in off-diagonal block).

Algorithm 5.4 Distributed matrix vector multiplication.

    function DISTR_MVMULT(A, x, y)
        Start asynchronous communication:
        Exchange ghost values;
        $y_{\text{int}} = A_{\text{diag}} \, x_{\text{int}}$;
        Synchronize communication;
        $y_{\text{int}} = y_{\text{int}} + A_{\text{offdiag}} \, x_{\text{ghost}}$;
    end function

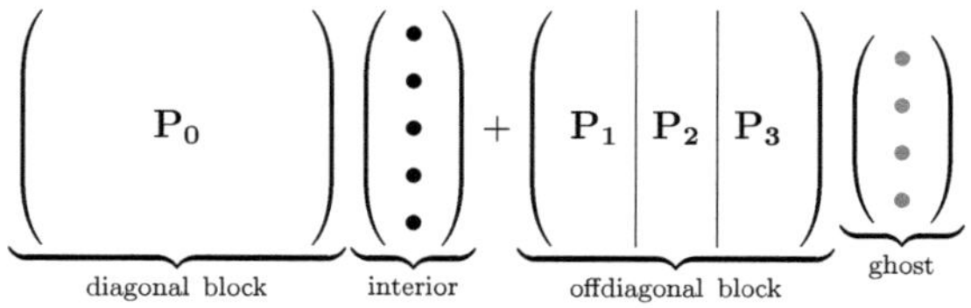

$$\underbrace{\begin{pmatrix} & & \\ & \mathbf{P_0} & \\ & & \end{pmatrix}}_{\text{diagonal block}} \underbrace{\begin{pmatrix} \bullet \\ \vdots \\ \bullet \end{pmatrix}}_{\text{interior}} + \underbrace{\begin{pmatrix} \mathbf{P_1} & \mathbf{P_2} & \mathbf{P_3} \end{pmatrix}}_{\text{offdiagonal block}} \underbrace{\begin{pmatrix} \bullet \\ \vdots \\ \bullet \end{pmatrix}}_{\text{ghost}}$$

Figure 5.2: Distributed matrix vector multiplication.

5.3. Parallelization

For the computation of pseudospectra, we need multiplications of a transposed matrix with vectors as well. However, using a row-wise distribution results in an all to all communication of the vector to be multiplied, see [55]. Thus, in order to reduce communication, we store the matrix twice: once with a row-wise distribution and once with a column-wise distribution.

5.3.2 Parallel Computation of Pseudospectra

For different grid points z, the computation of singular values of $z - A$ can be performed completely independently which allows an easy way to parallelize the evaluations of $\|(z-A)^{-1}\|_2$. However, this implies that a copy of the matrix A is available to each process. In order to avoid high storage costs in the case of large matrices, we utilize a parallel linear algebra as well.

We use an approach which is referred to as *hybrid parallelism*. Therefore, we partition the domain of grid points in $\mathbb{C}$ into k subdomains of the same size. Then p processes are mapped to each of these subdomain of grid points building k *groups*, provided that we have $k \times p$ processes available. Within each group the system matrices and vectors are spread among the processes in order to perform a parallel computation of the smallest singular vector, see Figure 5.3.

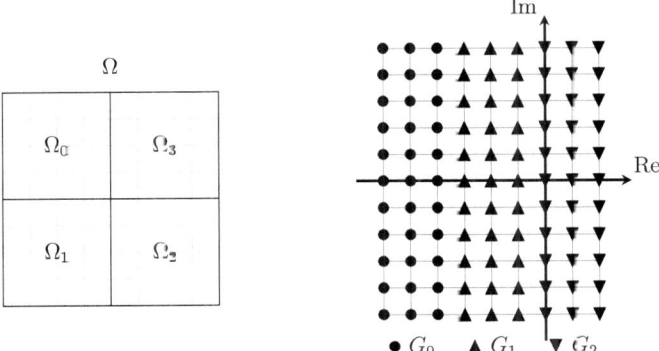

Figure 5.3: The left figure shows a decomposition of the physical domain Ω into four subdomains ($p = 4$). The right one depicts the distribution of grid points in $[x_1, x_2] \times [y_1, y_2] \subset \mathbb{C}$ among three groups G_i ($k = 3$).

Chapter 6

Pseudospectra for Applications in Hydrodynamic Stability

The succeeding numerical computations of spectral portraits were carried out with the predecessor of the finite element software *HiFlow^3*, see [1, 6]. In the two-dimensional case we have used the parallel direct sparse linear solver *MUMPS* as a preconditioner for the Davidson method. This solver is based on a multifrontal method, see [5]. Although this method may not be the best choice in terms of scalability, we have achieved the best performance results for two-dimensional problems. For three-dimensional problems, iterative solvers, such as the CG method, have been turned out to be successful as well. The parallel linear algebra employed in this work was implemented by means of the library *PETSc*, see [10, 11, 12].

The numerical benchmarks were performed on the *HP XC3000* cluster at the *Steinbuch Centre for Computing* (SCC), Karlsruhe, Germany. Each of the cores on the CPUs (Quad-core Intel Xeon processors) runs at a clock speed of 2.53 GHz.

6.1 Incompressible Flow over a Backward Facing Step

We consider a stationary incompressible fluid flow over a backward facing step as depicted in Figure 6.1. This set up is originated from a well-known optimization problem where the vortex behind the step is to be reduced, see e.g. [27]. The fluid flow in the domain Ω is governed by the steady-state Navier-Stokes equations

$$-\frac{1}{Re}\Delta \mathbf{v} + (\mathbf{v} \cdot \nabla)\mathbf{v} + \nabla p = \mathbf{0},$$
$$\nabla \cdot \mathbf{v} = 0, \quad (6.1)$$

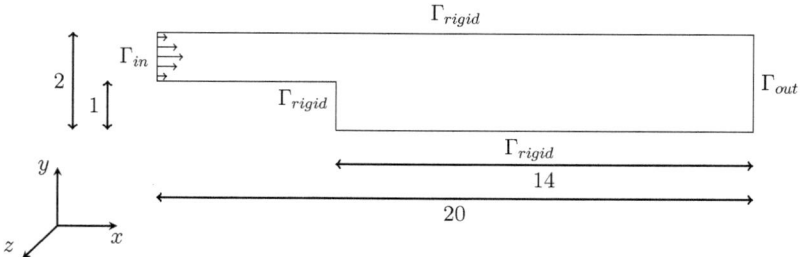

Figure 6.1: Geometry of the backward facing step benchmark.

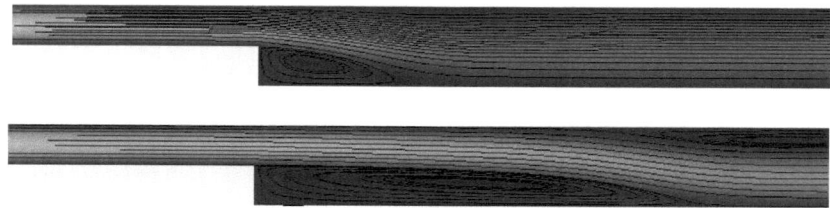

Figure 6.2: Stationary flow in the two-dimensional backward facing step geometry with $Re = 100$ (upper) and $Re = 500$ (lower).

where we have set $\rho = 1$. Here $Re = VL/\nu$ denotes the Reynolds number with characteristic velocity V and characteristic length L. As an inflow boundary condition we set

$$\mathbf{v} = \mathbf{v}_{in} \quad \text{on } \Gamma_{in},$$

where $\mathbf{v}_{in}$ has a parabolic profile. On Γ_{rigid} we impose *no-slip* boundary conditions, i.e.

$$\mathbf{v} = \mathbf{0} \quad \text{on } \Gamma_{rigid}.$$

Furthermore, we prescribe free-stream outflow conditions (or *do-nothing* conditions)

$$\frac{1}{Re}\partial_\mathbf{n}\mathbf{v} - p\,\mathbf{n} = 0 \quad \text{on } \Gamma_{out},$$

where $\mathbf{n}$ refers to the outward unit normal.

In our numerical examples we have chosen V to be the maximum velocity of the parabolic inflow $\mathbf{v}_{in}$, and L to be the gap height. Stationary solutions for $Re = 100$ and $Re = 500$ of the two-dimensional problem are plotted in Figure 6.2.

6.1. Incompressible Flow over a Backward Facing Step

We start with the variational formulation of the steady problem. We set

$$H(\Omega) = \{\mathbf{v} \in (H^1(\Omega))^d : \mathbf{v} = \mathbf{0} \text{ on } \Gamma_{in} \cup \Gamma_{rigid}\}$$

for $d = 2$ or $d = 3$. Moreover, let $\mathbf{v}_{in}^{ext} \in (H^1(\Omega))^d$ be a solenoidal extension of the inflow $\mathbf{v}_{in}$ such that $\mathbf{v}_{in}^{ext} = \mathbf{0}$ on Γ_{rigid}, see e.g. [47]. Then, the weak formulation seeks for pairs $(\mathbf{V}, P) \in H(\Omega) \times L^2(\Omega) + \{\mathbf{v}_{in}^{ext}, 0\}$ such that

$$\frac{1}{Re}(\nabla \mathbf{V}, \nabla \boldsymbol{\varphi})_0 + ((\mathbf{V} \cdot \nabla)\mathbf{V}, \boldsymbol{\varphi})_0 - (P, \nabla \cdot \boldsymbol{\varphi})_0 + (\nabla \cdot \mathbf{V}, \psi) = 0$$

holds for any $(\boldsymbol{\varphi}, \psi) \in H(\Omega) \times L^2(\Omega)$. We solve this equation numerically with a conforming Galerkin finite element method by replacing the spaces $H(\Omega)$ and $L^2(\Omega)$ by finite-dimensional subspaces H_h and L_h. Hence, we have to find $\{\mathbf{V}_h, P_h\} \in H_h \times L_h + \{\mathbf{v}_{in,h}^{ext}, 0\}$ satisfying

$$\frac{1}{Re}(\nabla \mathbf{V}_h, \nabla \boldsymbol{\varphi}_h)_0 - ((\mathbf{V}_h \cdot \nabla)\mathbf{V}_h, \boldsymbol{\varphi}_h)_0 - (P_h, \nabla \cdot \boldsymbol{\varphi}_h)_0 + (\nabla \cdot \mathbf{V}_h, \psi_h) = 0 \quad (6.2)$$

for any $(\boldsymbol{\varphi}, \psi) \in H_h \times L_h \subset H(\Omega) \times L^2(\Omega)$. Here $\mathbf{v}_{in,h}^{ext}$ is an adequate boundary interpolant of $\mathbf{v}_{in}^{ext}$, see [91].

We choose a triangulation $\mathcal{T}_h$ of Ω consisting of quadrilaterals (for $d = 2$) or hexahedrons (for $d = 3$) such that the obtained mesh is *affine*, i.e. each $K \in \mathcal{T}_h$ is affine equivalent to the reference element $\hat{K} := (0,1)^d$. This means for any $K \in \mathcal{T}_h$ there exists an affine and orientation preserving mapping F_K such that $K = F_K(\hat{K})$, see e.g. [41].

We consider so-called Q_k *finite elements* which consist of continuous, piecewise polynomial functions of degree k, see e.g. [41]. Let

$$Q_k(\hat{K}) = \text{span}\{x^i y^j : 0 \leq i, j \leq k\}, \quad \text{for } d = 2,$$
$$Q_k(\hat{K}) = \text{span}\{x^i y^j z^l : 0 \leq i, j, l \leq k\}, \quad \text{for } d = 3,$$

be the polynomials up to the order k on the reference cell $\hat{K}$. Then, the vector space of Q_k elements is given by

$$X_h^k := \{u_h \in C(\overline{\Omega}) : u_h|_K \circ F_K \in Q_k(\hat{K}) \text{ for any } K \in \mathcal{T}_h\}.$$

In our numerical experiments we have chosen Q_2 elements for the velocity and Q_1 elements for the pressure, i.e. $H_h = \{\mathbf{v}_h \in (X_h^2)^d : \mathbf{v}_h = \mathbf{0} \text{ on } \Gamma_{in} \cup \Gamma_{rigid}\}$ and $L_h = X_h^1$,

Algorithm 6.1 Solution procedure in HiFlow.
1: Read initial mesh and perform h-refinement;
2: Define FEM on the cells (e.g. Q_1 or Q_2);
3: Solve (6.2) by means of a Newton method;
4: Assemble $\mathbf{A}_h$ and $\mathbf{M}_h$;
5: **for** μ on a grid in $\mathbb{C}$ **do**
6:     Compute smallest singular value ς_h by means of (6.5) with the Davidson method;
7: **end for**
8: Plot spectral portrait;

in order to satisfy the *inf-sup-condition*

$$\inf_{\psi \in L_h} \sup_{\varphi \in H_h} \frac{|(\nabla \cdot \varphi, \psi)_0|}{\|\varphi\|_{H^1(\Omega)} \|\psi\|_{L^2(\Omega)}} > 0.$$

For pairs $\mathbf{u}_h = \{\mathbf{v}_h, p_h\}$ and $\boldsymbol{\xi}_h = \{\varphi_h, \psi_h\} \in H_h \times L_h$ we define the sesquilinear forms

$$a(\mathbf{u}_h, \boldsymbol{\xi}_h) = \frac{1}{Re}(\nabla \mathbf{v}_h, \nabla \varphi_h)_0 + ((\mathbf{v}_h \cdot \nabla)\mathbf{V}_h, \varphi_h)_0 + ((\mathbf{V}_h \cdot \nabla)\mathbf{v}_h, \varphi_h)_0$$
$$- (p_h, \nabla \cdot \varphi_h)_0 + (\nabla \cdot \mathbf{v}_h, \psi_h)_0,$$
$$m(\mathbf{u}_h, \boldsymbol{\xi}_h) = (\mathbf{v}_h, \varphi_h)_0,$$

where we have derived $a(\cdot, \cdot)$ just by linearization of (6.2) around a steady solution $\{\mathbf{V}_h, P_h\}$ as described in Section 2.2. Let $\boldsymbol{\xi}_h^i$ ($i = 1, \ldots, n$) be a basis of the finite-dimensional space $H_h \times L_h$. We denote by $\mathbf{A}_h$ the *stiffness matrix*

$$\mathbf{A}_h = \left(a(\boldsymbol{\xi}_h^j, \boldsymbol{\xi}_h^i)\right)_{1 \leq i,j \leq n}, \tag{6.3}$$

and by $\mathbf{M}_h$ the symmetric positive semidefinite *mass matrix*

$$\mathbf{M}_h = \left(m(\boldsymbol{\xi}_h^j, \boldsymbol{\xi}_h^i)\right)_{1 \leq i,j \leq n}. \tag{6.4}$$

The smallest singular values ς_h are determined by finding the smallest eigenvalues ς_h^2 and the corresponding eigenvectors x_h of

$$(\mu \mathbf{M}_h - \mathbf{A}_h)^H (\mu \mathbf{M}_h - \mathbf{A}_h) x_h = \varsigma_h^2 \mathbf{M}_h^H \mathbf{M}_h x_h, \tag{6.5}$$

for some μ on a grid in the complex plane. The complete solution procedure is outlined in Algorithm 6.1.

Spectral portraits of the two-dimensional problem are plotted in Figure 6.3 with the Reynolds number ranging from 100 to 500 as indicated. Here, we computed singular values for 4,488 grid points in the complex domain $[-0.6, 0.2] \times [0, 0.4] \subset \mathbb{C}$ and exploited

6.1. Incompressible Flow over a Backward Facing Step

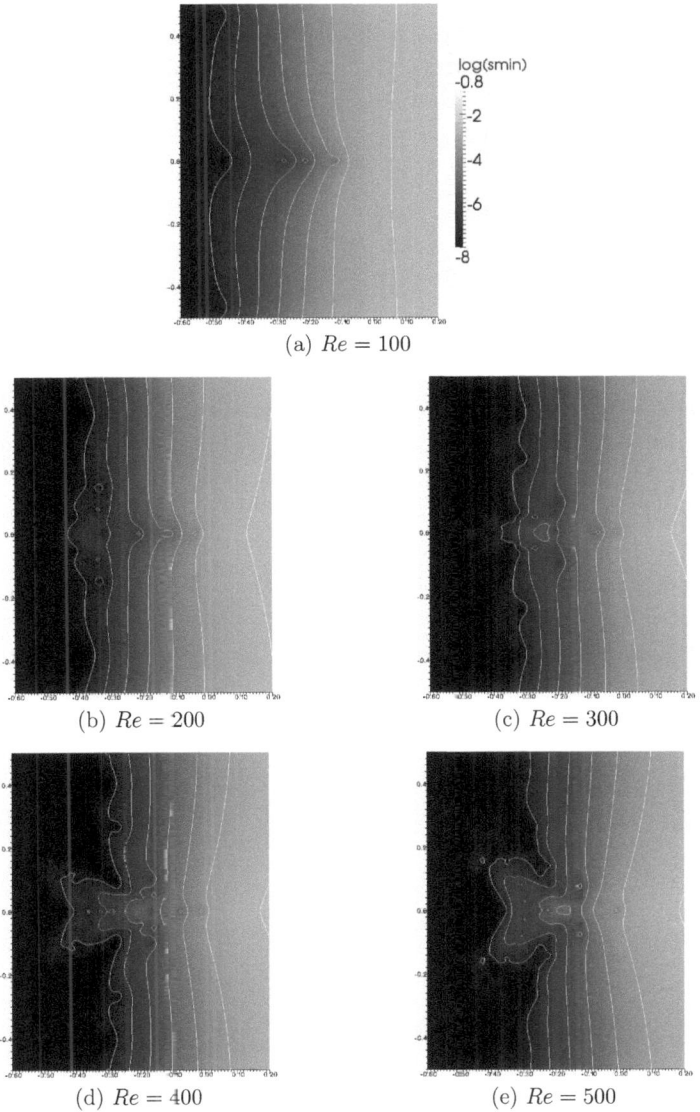

Figure 6.3: Spectral portraits in the domain $[-0.6, 0.2] \times [-0.4, 0.4]$ of the two-dimensional problem for different Reynolds numbers with contour plots for $\{-7, -6, ..., -1\}$.

the symmetry along the real axis. The 259,971 unknowns were distributed among 4 cores and the computation time for each of the depicted spectral portraits took around $110-135$ minutes on 4×8 cores on the HP XC3000 cluster described above. For increasing Reynolds numbers the pseudospectra protrude more and more into the right half of the complex plane which may indicate instability.

In order to test strong scalability of our computations we have set up a two-dimensional benchmark problem choosing $Re = 100$. This benchmark consists of the computation of 672 singular values on a grid in $[-0.4, 0.2] \times [0, 0.4] \subset \mathbb{C}$. The $259,971$ unknowns were distributed among 2^i, $i = 2, \ldots, 8$ cores exploiting the hybrid parallelism as described in Section 5.3.2. In Table 6.1 the results in terms of strong scalability using one group are given. In Table 6.2 we find the corresponding results in case of two groups. By distributing the unknowns over 4 processes, we have obtained the best performance, see Table 6.3. Here, we spread the grid points, divided in vertical stripes, equally among each group as in Figure 5.3. Altough this seems to be a quite simple load balancing technique, we have achieved a significant improvement, see Figure 6.4.

Solutions of the steady flow in the three-dimensional case for $Re = 100$ and $Re = 500$ are plotted in Figure 6.5. As in the two-dimensional case, the vortex behind the backward facing step enlarges as the Reynolds number increases. This tendency to instability is reflected by the pseudospectra, which are protruding more and more into the right half of the complex plane, see Figure 6.7. Here, the discretization of the problem yielded 143,484 unknowns, which were spread among 8 cores. We have chosen the same setup as in the two-dimensional case, i.e. we plotted singular values for 4,488 grid points in $[-0.6, 0.2] \times [0, 0.4] \subset \mathbb{C}$.

The estimate (2.30) in Theorem 2.13 states a minimum growth factor for the evolution of the perturbed quantity by means of pseudospectra: For any $a = \operatorname{Re} z > 0$ and $L = a \, \|(z-A)^{-1}\|$ we have

$$\sup_{0 < t \leq \tau} \|e^{tA}\| \geq e^{\tau a} \bigg/ \left(1 + \frac{e^{\tau a} - 1}{L}\right). \tag{6.6}$$

For the three-dimensional problem we have plotted this bound in Figure 6.6, where we have chosen a within the complex domain shown in Figure 6.7 such that L attains its maximum value. The plot shows that for increasing Reynolds numbers, the lower bound increases as well. This coincides with the observation that a steady flow tends to instability for increasing Reynolds numbers.

6.1. Incompressible Flow over a Backward Facing Step

group size	number cores	time (s)	speedup	efficiency
4	4	6653	1.	1.
8	8	5185	1.28	0.64
16	16	4266	1.56	0.39
32	32	2889	2.30	0.29
64	64	2788	2.39	0.15

Table 6.1: Scalability test using one group.

group size	number cores	time (s)	speedup	efficiency
4	8	4313	1.	1.
8	16	3162	1.36	0.68
16	32	2513	1.72	0.43
32	64	1705	2.53	0.32
64	128	1628	2.65	0.17

Table 6.2: Scalability test using two groups.

number cores	time (s)	speedup	efficiency
4	6653	1.	1.
8	4313	1.54	0.77
16	2447	2.72	0.68
32	1301	5.11	0.64
64	691	9.63	0.60
128	358	18.57	0.58
256	177	37.50	0.59

Table 6.3: Scalability test with group size four.

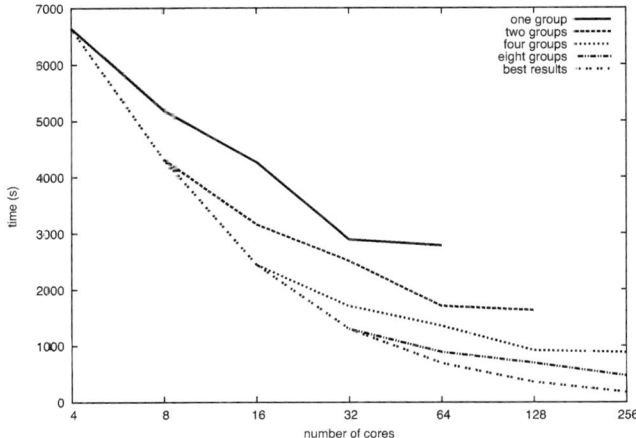

Figure 6.4: Running times for different group setups.

Figure 6.5: Stationary flow in the three-dimensional backward facing step geometry with $Re = 100$ (upper) and $Re = 500$ (lower).

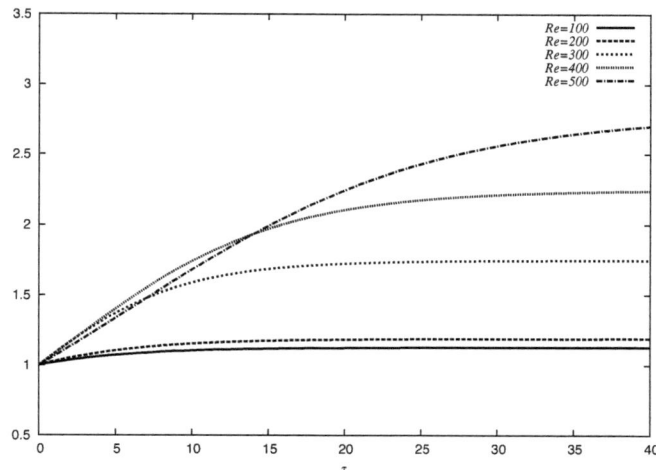

Figure 6.6: Plot of lower bound (6.6) with respect to the three-dimensional problem with Reynolds numbers ranging from 100 to 500 as indicated.

6.1. Incompressible Flow over a Backward Facing Step

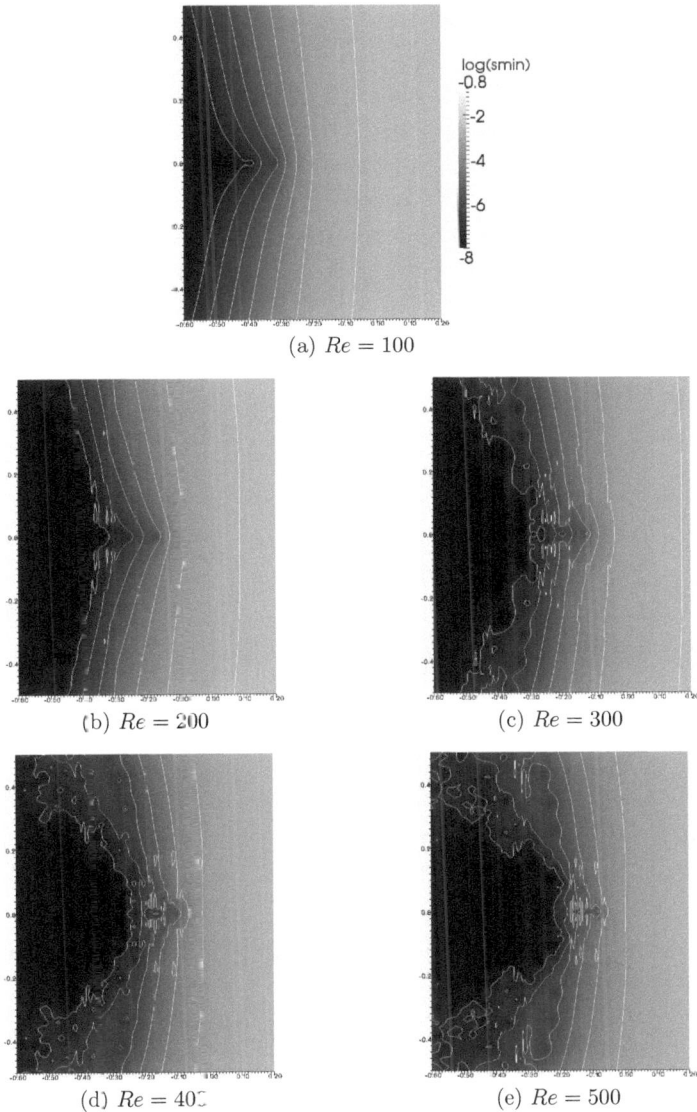

Figure 6.7: Spectral portraits in the domain $[-0.6, 0.2] \times [-0.4, 0.4]$ of the three-dimensional problem for different Reynolds numbers with contour plots for $\{-7, -6, ..., -1\}$.

6.2 Natural Convection in a Horizontal Annulus

In this section we consider natural convective flows of Newtonian fluids in an infinitely long gap between two horizontal coaxial cylinders. This physical process appears in a variety of applications, such as designing energy storage systems, cooling of electronic components, or modeling aircraft cabin insulations.

We denote the radius of the inner cylinder by R_i, and by R_o the radius of the outer cylinder (thus $0 < R_i < R_0$). Moreover, we prescribe constant surface temperatures T_i on the inner and T_o on the outer jacket with $T_o < T_i$. Modeling this system with the Oberbeck-Boussinseq equations (2.8), the flow pattern is characterized by three dimensionless quantities. We have the geometric parameter

$$\mathcal{A} = \frac{2R_i}{R_o - R_i}$$

which is known as *inverse relative gap width*. Moreover, Ra denotes the *Rayleigh number* defined by

$$Ra = \frac{\alpha g}{\nu \kappa}(T_i - T_o)(R_o - R_i)^3,$$

with the volumetric expansion coefficient α, gravity acceleration g, kinematic viscosity ν, and thermal diffusity κ. Finally, the *Prandtl number* is given by

$$Pr = \frac{\nu}{\kappa}.$$

Depending on these characteristics, the problem is very complex and allows very different possible behavior of the flow. Up to now, it was mostly investigated by numerical experiments, see e.g. [118, 119] and references therein, but there exist some theoretical results as well, see [77, 78] and references therein.

Despite the fact that spiral flows may occur, we confine ourselves to a two-dimensional model, which is physically reasonable for low Rayleigh numbers as pointed out in [39]. Hence, the domain under consideration is given by

$$\Omega = \{(r, \varphi) \colon R_i < r < R_o\, , 0 \leq \varphi < 2\pi\},$$

see Figure 6.8.

As described in [117], if the Rayleigh number exceeds a critical value, two steady solutions are observed: The *downward flow* consists of two counter rotating vortices in each half of the annulus, whereas the *upward flow* has a crescent shaped eddy on each side, see Figure 6.9.

6.2. Natural Convection in a Horizontal Annulus

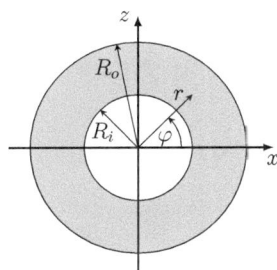

Figure 6.8: Two-dimensional model of the flow region.

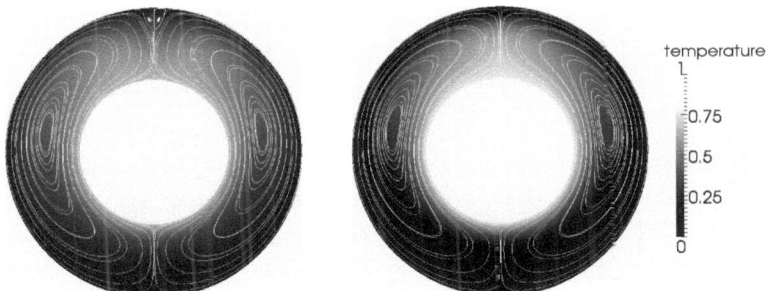

Figure 6.9: Streamlines and temperature distribution of the downward flow (left) and upward flow (right) for $\mathcal{A} = 2$, $Pr = 0.7$, $Ra = 5{,}000$.

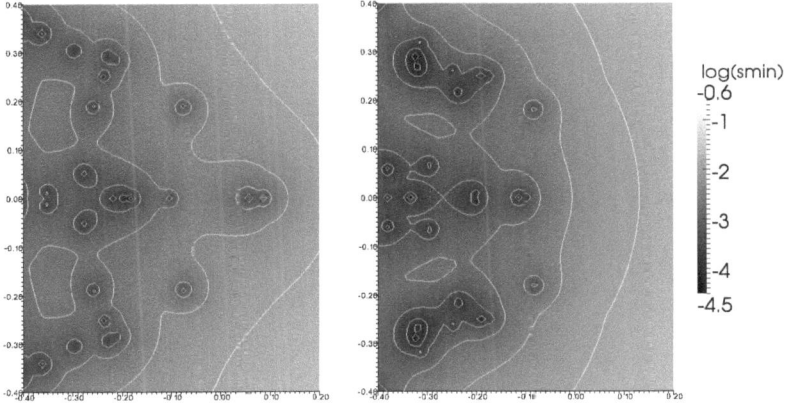

Figure 6.10: Spectral portraits of the downward flow (left) and the upward flow (right) in the domain $[-0.4, 0.2] \times [-0.4, 0.4]$ with contour plots for $\{-4, -3.5, \ldots, -1\}$.

The governing steady-state Oberbeck-Boussinseq equations read

$$-\nu \Delta \mathbf{v} + (\mathbf{v} \cdot \nabla)\mathbf{v} + \frac{1}{\rho}\nabla p = \mathbf{g}\left[1 + \alpha(T_0 - T)\right],$$

$$\nabla \cdot \mathbf{v} = 0,$$

$$\partial_t T + \mathbf{v} \cdot \nabla T - \kappa \Delta T = 0,$$

see Section 2.1.6. We impose no-slip boundary conditions for the velocity, i.e.

$$\mathbf{v} = \mathbf{0} \quad \text{on } \partial \Omega.$$

Moreover, we set $T_i = 1$ and $T_o = 0$ which results in the following boundary conditions for the temperature:

$$T(\mathbf{x}) = 1 \quad \text{for } \|\mathbf{x}\|_2 = R_i,$$
$$T(\mathbf{x}) = 0 \quad \text{for } \|\mathbf{x}\|_2 = R_o.$$

For the discretization we choose Q_2 elements for the velocity and temperature and Q_1 elements for the pressure. Thus, with the notations $H_h = \{\mathbf{v}_h \in (X_h^2)^2 : \mathbf{v}_h = \mathbf{0} \text{ on } \partial\Omega\}$, $L_h = \{p_h \in X_h^1 : \int_\Omega p_h = 0\}$ and $W_h = \{T_h \in X_h^2 : T_h = 0 \text{ on } \partial\Omega\}$ and an appropriate boundary interpolant $T_{i,h}^{ext}$ of the imposed temperature on the inlet, we seek for $(\mathbf{V}_h, P_h, T_h) \in H_h \times L_h \times W_h + \{\mathbf{0}, 0, T_{i,h}^{ext}\}$ such that

$$\begin{aligned}
&\nu(\nabla \mathbf{V}_h, \nabla \boldsymbol{\varphi}_h)_0 + ((\mathbf{V}_h \cdot \nabla)\mathbf{V}_h, \boldsymbol{\varphi}_h)_0 - (P_h, \nabla \cdot \boldsymbol{\varphi}_h)_0 + \mathbf{g}\alpha(T_h, \boldsymbol{\varphi}_h)_0 \\
&+ (\nabla \cdot \mathbf{V}_h, \psi_h)_0 \\
&+ (\mathbf{V}_h \cdot \nabla T_h, \zeta)_0 + \kappa(\nabla T_h, \nabla \zeta_h)_0 \\
&= \alpha T_o(\mathbf{g}, \boldsymbol{\varphi}_h)_0
\end{aligned} \tag{6.7}$$

holds for any $(\boldsymbol{\varphi}_h, \psi_h, \zeta_h) \in H_h \times L_h \times W_h$. For brevity, we have set $\rho = 1$. By linearization around a steady solution as in the preceding Section 6.1, we obtain for triples $\mathbf{u}_h = \{\mathbf{v}_h, p_h, T_h\}$ and $\boldsymbol{\xi}_h = \{\boldsymbol{\varphi}_h, \psi_h, \zeta_h\} \in (X_h^2)^2 \times X_h^1 \times X_h^2$ the sesquilinear form

$$\begin{aligned}
a(\mathbf{u}_h, \boldsymbol{\xi}_h) = &\ \nu(\nabla \mathbf{v}_h, \nabla \boldsymbol{\varphi}_h)_0 + ((\mathbf{v}_h \cdot \nabla)\mathbf{V}_h, \boldsymbol{\varphi}_h)_0 + ((\mathbf{V}_h \cdot \nabla)\mathbf{v}_h, \boldsymbol{\varphi}_h)_0 \\
&- (p_h, \nabla \cdot \boldsymbol{\varphi}_h)_0 + \mathbf{g}\alpha(T_h, \boldsymbol{\varphi}_h)_0 + (\nabla \cdot \mathbf{v}_h, \psi_h)_0 \\
&+ (\mathbf{v}_h \cdot \nabla T_h, \zeta_h)_0 + (\mathbf{V}_h \cdot \nabla T_h, \zeta_h)_0 + \kappa(\nabla T_h, \nabla \zeta_h)_0.
\end{aligned}$$

Moreover, we define

$$m(\mathbf{u}_h, \boldsymbol{\xi}_h) = (\mathbf{v}_h, \boldsymbol{\varphi}_h)_0 + (T_h, \zeta_h)_0.$$

6.2. Natural Convection in a Horizontal Annulus

Then, using the definitions for the stiffness matrix and the mass matrix as in (6.3) and (6.4), we obtain the same formulation as before: We seek the smallest eigenvalue ς^2 of the generalized eigenvalue problem

$$(\mu \mathbf{M}_h - \mathbf{A}_h)^H (\mu \mathbf{M}_h - \mathbf{A}_h) x_h = \varsigma_h^2 \, \mathbf{M}_h^H \mathbf{M}_h x_h.$$

Spectral portraits with respect to the steady solutions depicted in Figure 6.9 are plotted in Figure 6.10, where we have set $\mathcal{A} = 2$, $Pr = 0.7$, $Ra = 5{,}000$. For each of these plots we have chosen a discretization resulting in 429,568 unknowns and evaluated 2,624 singular values exploiting the symmetry along the real axis.

The pseudospectra of the downward flow are protruding more into the right half of the complex plane than the pseudospectra of the upward flow, which may indicate the downward flow to be closer to instability than the upward flow. However, as for the Bénard problem, if two steady solutions are realized, it is expected the upward flow to be instable and the downward flow to be stable. If the Reynolds number is increased any further, the downward flow is observed to become instable as well.

Chapter 7

Applications in Nuclear Reactor Theory

Up to now, the power method (cf. Algorithm 5.1) has been the method of choice for solving the k eigenvalue criticality problem (3.4) in reactor applications, see [66]. In order to find a better alternative to this rather basic method, several ways have been investigated: The employment of the implicit restarted Arnoldi method is examined in [22, 111, 112]. The Jacobi-Davidson method in this context is discussed in [52, 110], and, more recently, the application of the Jacobian-free Newton-Krylov method is examined in [45, 46, 60]. These techniques have been applied to the diffusion equation (i.e. P_1 approximation for the spherical dependence) or to the discrete ordinate transport approximation (i.e. S_N approximation) so far.

In this chapter the Davidson method (cf. Algorithm 5.2) is compared to the traditional power method in the framework of the P_N approximation. The numerical evaluations are performed within the framework of the software Parafish, see [107]. This parallel solver is based on a non-overlapping domain-decomposition (DD) technique. It uses the multigroup approach for the energy discretization, non-conforming finite elements for the spatial discretization, and spherical harmonics for the angular discretization as described in Section 4.2.

The order of discretization in Parafish is as follows: first the energy, then the space, and finally the angle. At the energy level, we use a block version of the Gauss-Seidel method given the block wise representation of the multigroup discretization, see Section 4.2.1. If no *upscattering* is present, i.e. $\sigma_s^{g' \to g} \equiv 0$ for $g < g'$, block triangularity is fully achieved and only one Gauss-Seidel iteration is sufficient.

At the spatio-angular level, domain-decomposition is applied yielding an interior block shape such that each interior diagonal block is symmetric and corresponds to one domain, see Section 4.2.2. Therefore, these diagonal blocks can be treated independently by different processes. At this spatio-angular level, we apply a block-diagonally preconditioned version of the GMRES method by factorizing each interior diagonal block by an incom-

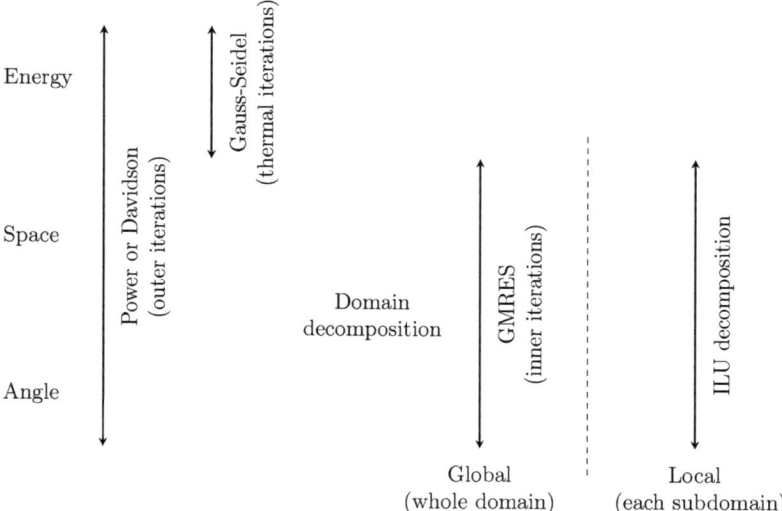

Figure 7.1: Parafish calculation scheme, see [107].

plete LU (ILU) factorization. The ILU decomposition was implemented by means of the libraries *SparseLib++* [36] and *SuperLU* [33]. The different iterations loops of Parafish are schematically displayed in Figure 7.1.

Moreover, Parafish is designed in such a way that one process can handle more than one domain. In this way, one can take full advantage of the flexibility of the resulting preconditioning without increasing the number of processes. Furthermore, Parafish allows the user to use different factorization methods in different domains. Note that the parallelization scheme used in Parafish differs from the one presented in Section 5.3. The interface nodes are duplicated among neighboring domains, see [107].

7.1 The Takeda 1 Benchmark

The three-dimensional Takeda 1 benchmark [100] consists of a small light water reactor (LWR) core of cubic shape as depicted in Figure 7.2. All units of length are given in cm. This corresponds to the simplified model of a reactor as shown in Figure 3.1 with one core and two control rods, and assuming symmetry along the planes $x = 0$, $y = 0$ and $z = 0$ (if we set the origin in the center of the reactor). The Takeda 1 benchmark considers only one octant of the reactor describing the remaining ones by reflecting boundary conditions.

7.1. The Takeda 1 Benchmark

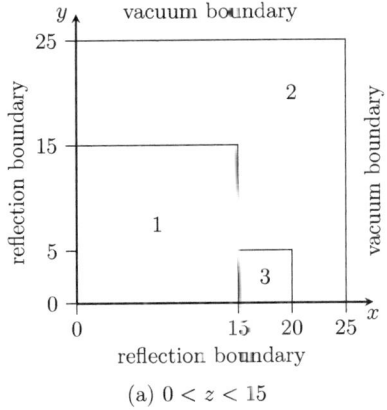
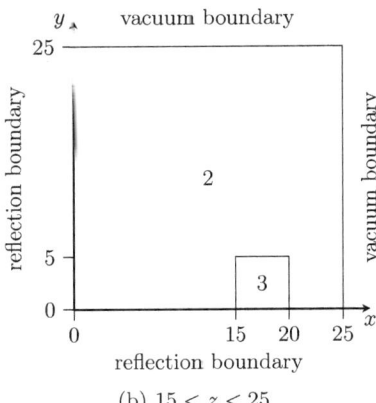

Figure 7.2: Geometry of the 3D Takeda 1 benchmark. Zone 1 = core, zone 2 = moderator, zone 3 = guide tube.

Hence, it prescribes reflective boundary conditions for $x = 0$, $y = 0$ and $z = 0$. On the outer frame, i.e. for $x = 25$ $y = 25$ and $z = 25$, we impose vacuum boundary conditions. Furthermore, we have two energy groups, where no upscattering is present, i.e. $\sigma_s^{2\to 1} \equiv 0$. The corresponding cross section values can be found in [100].

In the geometry three different zones are to be found: the core zone, the moderator zone and the guide tube zone. We consider both instances of this benchmark:

- Case 1: The control rod is in the guide tube zone.

- Case 2: The control rod is out of the core, i.e. the guide tube zone is voided.

For the angular discretization we use the P_N method, with N ranging from 3 to 7 as indicated. The spatial dependence is discretized by means of a $25 \times 25 \times 25$ mesh with hexahedral NC_6 elements. This element is a three-dimensional generalization of the two-dimensional rotated Q_1 element, see [83]. Its capability to accurately solve the Takeda 1 benchmark has been assessed previously in [106]. The finite element nodes are distributed among $5 \times 5 \times 5$ domains, with domain interfaces parallel to the cube faces. Note that this domain decomposition implies the guide tube zone to be entirely comprised within one single domain.

In case 1 of the benchmark, we have optimized our computational time by applying an ILU(0) factorization in all domains as a preconditioner for the GMRES method. In case 2, we have achieved the best results by applying a variant of an ILUTP (incomplete LU factorization with threshold and pivoting) method as described in [67] in the voided guide tube zone and an ILU(0) decomposition in all remaining domains. The results we obtained in [96, 97] are given in the Tables 7.1, 7.2 and 7.3. These computations were performed on the HP XC3000 cluster as described in Chapter 6. Please note that we listed only the times for solving the eigenvalue problem including the time for the ILU decomposition. We excluded the time for assembling the matrices.

For case 1, we obtained for the P_3 (P_5 and P_7) approximation the critical eigenvalue $k_{eff} = 0.96122$ (0.96222 and 0.96241 respectively) which agrees decently with the reference values of [99], e.g. $k_{eff} = 0.9623 \pm 0.00048$ with the Monte-Carlo method. Satisfying results have also been achieved in case 2, where our numerical experiments yielded for the P_3 (P_5 and P_7) approximation the critical eigenvalue $k_{eff} = 0.97247$ (0.97553 and 0.97633 respectively) which is comparable with the reference value 0.9778 ± 0.00046 obtained with the Monte-Carlo method in [99].

The computation of the critical eigenvalue and the corresponding eigenfunction using the Davidson method consumes roughly half of the time which is needed by applying the power method to gain comparable results. No improvement to the power method results could be made by utilizing a Chebyshev acceleration. According to the number of outer iterations, we have obtained similar results as shown in Table 7.4. With regards to the speedup, both the Davidson and the power method show a comparable behavior. This is quite evident since in both methods the same basic linear algebra routines are utilized.

The Davidson method appears to be a robust method to compute eigenvalues and the corresponding eigenvectors of a generalized eigenvalue problem since the residuals of the form $Fv - \lambda Hv$ are computed accurately (in exact arithmetic). Moreover, the Davidson method builds a full basis of an adequate projection space, whereas the power method projects only onto the space spanned by the last built vector of the form $(H^{-1}F)^j v$. Hence, the Davidson method keeps track of more information and therefore needs less iterations. Both methods require solving linear systems of the form $Hx = y$. The Davidson method turned out to be more robust in practice in the sense that the choice of the preconditioner is less restrictive than in the power method. This means, solving linear systems of the form $Hx = y$ can be done very roughly, while the power method needs a more accurate solving in order to converge.

Comparing the two cases of the Takeda 1 benchmark, we notice that the performance and scalability achieved in case 2 (rod out) cannot keep up with the results obtained in case 1 (rod in). Having the rod out of the core is the tougher case because of the low-density region appearing in the voided guide tube. As noticed before, the voided guide tube is comprised within one domain, which implies that the corresponding workload within this domain is larger compared to the case where the rod is in the core.

7.1. The Takeda 1 Benchmark

Case 1: rod in

	Davidson Method			Power Method		
Number Cores	Time (s)	Speedup	Efficiency	Time (s)	Speedup	Efficiency
1	21.8	1.	1.	41.0	1.	1.
5	5.3	3.88	0.78	11.2	3.66	0.73
25	1.6	14.03	0.56	3.1	13.14	0.53
125	1.1	20.02	0.16	1.2	33.20	0.27

Case 2: rod out

	Davidson Method			Power Method		
Number Cores	Time (s)	Speedup	Efficiency	Time (s)	Speedup	Efficiency
1	32.7	1.	1.	58.0	1.	1.
5	8.4	3.9	0.78	15.4	3.8	0.75
25	3.3	9.8	0.39	5.8	10.0	0.40
125	2.2	15.1	0.12	3.6	16.0	0.13

Table 7.1: Results with P_3 approximation (675,000 unknowns).

Case 1: rod in

	Davidson Method			Power Method		
Number Cores	Time (s)	Speedup	Efficiency	Time (s)	Speedup	Efficiency
1	98.5	1.	1.	182.8	1.	1.
5	24.1	4.08	0.82	46.67	3.92	0.78
25	6.9	14.23	0.57	12.1	15.10	0.60
125	2.4	41.25	0.33	4.0	45.53	0.36

Case 2: rod out

	Davidson Method			Power Method		
Number Cores	Time (s)	Speedup	Efficiency	Time (s)	Speedup	Efficiency
1	152.7	1.	1.	307.2	1.	1.
5	38.2	4.1	0.80	78.8	3.9	0.78
25	17.3	8.8	0.35	30.5	10.1	0.40
125	12.6	12.1	0.10	20.4	15.0	0.12

Table 7.2: Results with P_5 approximation (1,687,500 unknowns).

Case 1: rod in

Number Cores	Davidson Method			Power Method		
	Time (s)	Speedup	Efficiency	Time (s)	Speedup	Efficiency
1	239.8	1.	1.	479.7	1.	1.
5	69.8	3.43	0.69	141.0	3.40	0.68
25	19.2	12.46	0.50	35.1	13.68	0.55
125	5.0	47.57	0.38	9.5	50.67	0.41

Case 2: rod out

Number Cores	Davidson Method			Power Method		
	Time (s)	Speedup	Efficiency	Time (s)	Speedup	Efficiency
1	633.1	1.	1.	1117.6	1.	1.
5	161.6	3.9	0.78	268.4	4.2	0.83
25	90.7	7.0	0.28	139.0	8.0	0.32
125	77.8	8.1	0.07	112.0	10.0	0.08

Table 7.3: Results with P_7 approximation (3,150,000 unknowns).

Case 1: rod in

Angular Approximation	Davidson Method		Power Method	
	Number Outers	Number Inners	Number Outers	Number Inners
P_3	6	217	12	549
P_5	6	237	13	617
P_7	5	194	14	671

Case 2: rod out

Angular Approximation	Davidson Method		Power Method	
	Number Outers	Number Inners	Number Outers	Number Inners
P_3	7	279	12	644
P_5	7	205	14	813
P_7	7	334	15	918

Table 7.4: Number of outer (Davidson or power) and inner (GMRES) iterations for the Takeda 1 benchmark.

7.2 The NEA C5G7 Benchmark

In this section we consider the two-dimensional MOX fuel assembly benchmark issued by the Nuclear Energy Agency (NEA). It uses C5 MOX fuel and 7 energy groups, hence its "C5G7" nickname. As depicted in Figure 7.3 (again, all units of length are in *cm*), its geometry is a quarter core containing four fuel assemblies and the surrounding moderator. Each fuel assembly is comprised of a 17×17 lattice of square pin cells. Each of these pin cells is made out of a cylindrical section surrounded by moderator. These cylindrical sections either contain a fuel-clad mix or constitute available space for the insertion of an absorber rod. The pin cell compositions are detailed in Figure 7.4. The corresponding cross section values can be found in [74]. Note that, unlike the Takeda 1 benchmark, the C5G7 benchmark includes upscattering.

The Parafish model uses a finite element mesh with quadrilateral rotated Q_1 elements such that each pin cell is discretized into 14×14 elements. This approximation of the cylindrical section by a Cartesian mesh was shown in [106] to be appropriate because it preserves not only the volume ratio between the two pin components, but also the "density" of the meshing in both these components. As for the angular discretization, we consider here the P_3 approximation.

We optimized the running times of our benchmarks by applying the ILUTP preconditioner in every domain. As for the decomposition, we obtained good results using a 20×20 DD. The grouping was done such that one or several pin-cells belong to one domain, and, within the assemblies, a domain limit is necessarily a pin-cell limit (the converse is not true).

The resulting 400 domains were spread among 32 to 400 cores such that a decent load balancing was obtained. This load balancing was performed semi-automatically, with some fine-tuning done "by hand".

	Davidson Method			Power Method		
Number Outers	19			37		
Number Inners	3,287			14,651		
Number Cores	Time (s)	Speedup	Efficiency	Time (s)	Speedup	Efficiency
32	161	1.	1.	528	1	1.
64	96	1.68	0.84	325	1.62	0.81
256	30	5.41	0.68	100	5.28	0.66
400	20	8.00	0.64	72	7.38	0.59

Table 7.5: Results for the 2D C5G7 benchmark with P_3 approximation (14,962,584 unknowns). The inner iterations refer to GMRES iterations in both cases. Speedups and efficiencies are computed with respect to the 32 core case.

92   APPLICATIONS IN NUCLEAR REACTOR THEORY

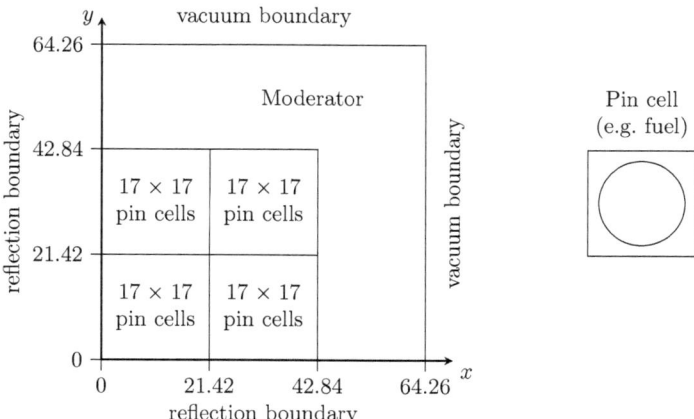

Figure 7.3: Geometry of the 2D C5G7 benchmark.

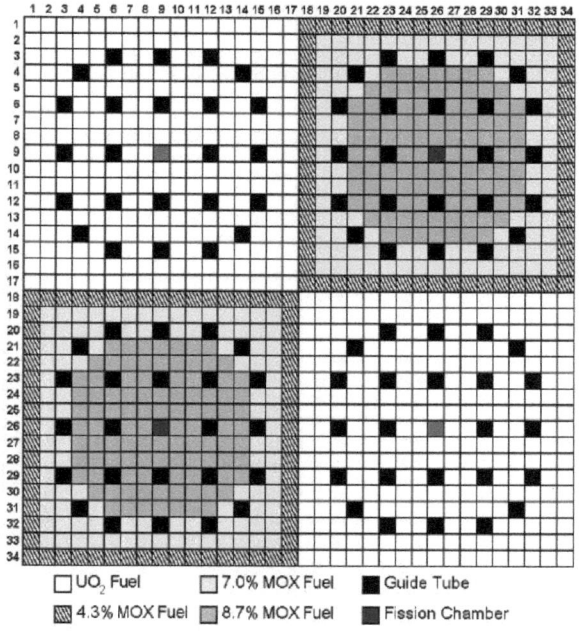

Figure 7.4: Pin cells composition in the C5G7 benchmark, see [74].

7.2. The NEA C5G7 Benchmark

The calculations for the C5G7 benchmark were run on the JuRoPA (Jülich Research on Petaflop Architectures) supercomputer installed at the Forschungszentrum Jülich, Germany. Each of the cores on the CPUs (Quad-core Intel Xeon processors) runs at a clock speed of 2.93 GHz.

We have in [74] a k effective reference value of $k_{eff} = 1.186550\pm0.00008$ obtained by the Monte-Carlo method. Both power and Davidson methods yield for the P_3 approximation a satisfying k effective of 1.18319 and 1.18320 respectively. The Davidson method was again much more efficient to reach this solution, as it can be seen in Table 7.5, cf. [96].

The computing times as well as the number of iterations are drastically reduced by using the Davidson method. Again, applying a Chebyshev acceleration to the power method could not improve our results.

Chapter 8
Summary and Outlook

In this book we have considered two involved classes of applications in terms of stability. We have illustrated that an efficient implementation exploiting parallel platforms enables us to address more and more complex problems.

With regards to the linear stability, pseudospectra have been shown to be an encouraging tool to facilitate the studies of nonlinear effects which may not be revealed by a traditional linear stability analysis. We have established a theoretical background to approximate spectral portraits of elliptic differential operators in terms of finite element methods. The numerical examples covered in this work include an incompressible fluid flow over a backward facing step as well as a natural convection process in an annulus. For both problems, we have illustrated that the Davidson method is capable to solve the inherent singular value problems efficiently.

As we have seen, pseudospectra protruding considerably into the right half of the complex plane may indicate a growth of perturbations. However, it still remains open to find a more accurate interpretation of pseudospectra. We have shown that a performant computation of pseudospectra on parallel platforms can be achieved. Nevertheless, a more sophisticated load balancing to further exploit the trivial parallelism can be developed in order to tune the scalability. For instance, one might include some dynamic assignment of regions in the complex plane dependent on the workload of the process groups. Furthermore, using parallel preconditioning techniques, such as multigrid or Schur complement methods, may turn out to be valuable to precondition the inherent eigenvalue computations.

As for the criticality problem, our numerical experiments have shown the superiority of the Davidson method in comparison with the so far widely used power method. This is fully comprehensible due to the fact that the Davidson method keeps track of more information along the iterative procedure. The extra storage costs and additional computational effort per iteration of the Davidson method are outweighted by the significant reduction of the number of iterations. Furthermore, the Davidson method appears to be more robust than the power method: First, the Davidson method controls the (mathematically) exact residuals of the matrix pencil. Second, it needs a less accurate solving in each preconditioning step than the power method. The power method applied to a matrix pencil may not need an exact solving of the arising linear systems in practice as it is required in theory, but it may not converge at all if this solving is done to roughly.

There is still potential to further tune the computation of critical eigenvalues which might be useful in the field of neutron transport theory in general. In particular, for problems including large upscattering, a coupled solving on the energy level instead of a block-wise applied Gauss-Seidel method may turn out to be valuable. Furthermore, especially for large-scaled problems, a parallelization scheme for the energy dependence can be introduced to improve scalability.

Appendix A

Estimates and Calculations for the Poincaré Constant in an Annulus

In this chapter we derive a bound for the Poincaré constant in an annulus and perform numerical evaluations which give us an insight about the quality of this bound. We apply these results to make quantitative statements of the approximation quality of a simplified model describing natural convection. These results are published in [54].

A.1 Bounds for the Poincaré Constant

As already seen in Section 2.2, the Poincaré constant is a very important tool to investigate fluid flows. For instance, it is needed to prove existence of weak solutions and regularity assumptions in the context of elliptic differential equations, see e.g. [19]. But as seen in Section 2.2, it plays a major role in the stability theory as well.

Let $\Omega \subset \mathbb{R}^n$ be bounded in at least one direction. The *Poincaré constant* is defined as the smallest constant k_p such that

$$\|u\|_0 \leq k_p \|\nabla u\|_0 \tag{A.1}$$

holds for all $u \in H_0^1(\Omega)$. If Ω is bounded, one can show that $k_p = \lambda_1^{-1/2}$, where λ_1 is the smallest eigenvalue of the Laplace problem

$$-\Delta u = \lambda u \quad \text{in } \Omega, \quad u = 0 \quad \text{on } \partial\Omega, \tag{A.2}$$

see e.g. [95]. The following two bounds we establish hold for quite general domains Ω and can be found in [44].

ESTIMATES AND CALCULATIONS FOR THE POINCARÉ CONSTANT

Theorem A.1 *If* $\Omega \subset \{x \in \mathbb{R}^n : -d/2 < x_n < d/2\}$, *then we have*

$$\|u\|_0 \leq \frac{d}{\pi}\|\nabla u\|_0$$

for all $u \in H_0^1(\Omega)$.

Proof. Since $C_0^\infty(\Omega)$ is dense in $H_0^1(\Omega)$, it is sufficient to show the assertion holds for all $u \in C_0^\infty(\Omega)$. We define the function

$$U(x) = \frac{u(x)}{\sin[\pi(x_n + d/2)/d]}.$$

Clearly, U is bounded and vanishes at $-d/2$ and $d/2$. We have

$$0 \leq \int_{-d/2}^{d/2} \left\{ \frac{\partial u}{\partial x_n} - \frac{\pi}{d} u \cot\left[\frac{\pi}{d}(x_n + \frac{d}{2})\right] \right\}^2 dx_n$$

$$= \int_{-d/2}^{d/2} \left\{ \frac{\partial u}{\partial x_n} \right\}^2 dx_n - \frac{\pi}{d} \int_{-d/2}^{d/2} \left\{ 2\frac{\partial u}{\partial x_n} u \cot\left[\frac{\pi}{d}(x_n + \frac{d}{2})\right] \right\} dx_n$$

$$+ \frac{\pi^2}{d^2} \int_{-d/2}^{d/2} \left\{ u \cot\left[\frac{\pi}{d}(x_n + \frac{d}{2})\right] \right\}^2 dx_n.$$

Integrating the second term by parts yields

$$-\frac{\pi}{d} \int_{-d/2}^{d/2} \left\{ 2\frac{\partial u}{\partial x_n} u \cot\left[\frac{\pi}{d}(x_n + \frac{d}{2})\right] \right\} dx_n$$

$$= -\frac{\pi}{d} u^2 \cot\left[\frac{\pi}{d}(x_n + \frac{d}{2})\right] \Big|_{-d/2}^{d/2} + \frac{\pi}{d}\int_{-d/2}^{d/2} \left\{ u^2 \left(-\frac{\pi}{d}\sin^{-2}\left[\frac{\pi}{d}(x_n + \frac{d}{2})\right]\right) \right\} dx_n$$

$$= -\frac{\pi}{d} u \cos\left[\frac{\pi}{d}(x_n + \frac{d}{2})\right] U \Big|_{-d/2}^{d/2} - \frac{\pi^2}{d^2}\int_{-d/2}^{d/2} \left\{ u^2 \sin^{-2}\left[\frac{\pi}{d}(x_n + \frac{d}{2})\right] \right\} dx_n$$

$$= -\frac{\pi^2}{d^2} \int_{-d/2}^{d/2} \left\{ u^2 \sin^{-2}\left[\frac{\pi}{d}(x_n + \frac{d}{2})\right] \right\} dx_n.$$

Therefore, we have

$$0 \leq \int_{-d/2}^{d/2} \left\{ \frac{\partial u}{\partial x_n} \right\}^2 dx_n - \frac{\pi^2}{d^2} \int_{-d/2}^{d/2} u^2 \left\{ \sin^{-2}\left[\frac{\pi}{d}(x_n + \frac{d}{2})\right] - \cot\left[\frac{\pi}{d}(x_n + \frac{d}{2})\right] \right\} dx_n.$$

Since

$$\sin^{-2} x - \cot^2 x = \frac{1 + \cos^2 x}{\sin^2 x} = \frac{1 + 1 - \sin^2 x}{\sin^2 x} = \frac{2}{\sin^2 x} - 1 \geq 1,$$

A.1. Bounds for the Poincaré Constant

we obtain

$$\int_{-d/2}^{d/2} u^2 \, dx_n \leq \frac{d^2}{\pi^2} \int_{-d/2}^{d/2} \left(\frac{\partial u}{\partial x_n}\right)^2 dx_n.$$

Let H^{n-1} denote the $(n-1)$-dimensional hyperplane of the first $n-1$ dimensions, i.e. $H^{n-1} = \operatorname{span}\{x_1, x_2, \ldots, x_{n-1}\}$. Then, we have

$$\|u\|_0^2 = \int_{H^{n-1}} \int_{-d/2}^{d/2} u^2 \, dx_n \, d(x_1, \ldots, x_{n-1}) \leq \int_{H^{n-1}} \int_{-d/2}^{d/2} \frac{d^2}{\pi^2} \left(\frac{\partial u}{\partial x_n}\right)^2 dx_n \, d(x_1, \ldots, x_{n-1})$$

$$= \frac{d^2}{\pi^2} \int_\Omega \left(\frac{\partial u}{\partial x_n}\right)^2 dx \leq \frac{d^2}{\pi^2} \int_\Omega \sum_{i=1}^n \left(\frac{\partial u}{\partial x_i}\right)^2 dx = \frac{d^2}{\pi^2} \|\nabla u\|_0^2,$$

which completes the proof. □

Theorem A.2 *Let $\Omega \subset \{x \in \mathbb{R}^n : -d/2 < x_i < d/2, i = 1, \ldots, n\}$, then*

$$\|u\|_0 \leq \frac{d}{\pi\sqrt{n}} \|\nabla u\|_0 \tag{A.3}$$

for any $u \in H_0^1(\Omega)$.

Proof. Again, it is enough to show the theorem for any $u \in C_0^\infty(\Omega)$. From the proof of the last theorem we know that

$$\int_{-d/2}^{d/2} u^2 \, dx_n \leq \frac{d^2}{\pi^2} \int_{-d/2}^{d/2} \left(\frac{\partial u}{\partial x_i}\right)^2 dx_i$$

holds for any $i = 1, \ldots, n$. Denoting the $(n-1)$-dimensional hyperplane of the dimensions $(1, \ldots, i-1, i+1, \ldots, n)$ by H_i^{n-1}, we deduce

$$n\|u\|_0^2 = \sum_{i=1}^n \int_\Omega u^2 \, dx$$

$$= \sum_{i=1}^n \int_{H_i^{n-1}} \int_{-d/2}^{d/2} u^2 \, dx_i \, d(x_1, \ldots, x_{i-1}, x_{i+1}, \ldots, x_n)$$

$$\leq \sum_{i=1}^n \int_{H_i^{n-1}} \frac{d^2}{\pi^2} \int_{-d/2}^{d/2} \left(\frac{\partial u}{\partial x_i}\right)^2 dx_i \, d(x_1, \ldots, x_{i-1}, x_{i+1}, \ldots, x_n)$$

$$= \frac{d^2}{\pi^2} \int_\Omega \sum_{i=1}^n \left(\frac{\partial u}{\partial x_i}\right)^2 dx = \frac{d^2}{\pi^2} \|\nabla u\|_0^2,$$

which completes the proof. □

From the last theorem, we obtain in terms of the annulus $\Omega_A = \{(r, \varphi) : A/2 < r <$

$1 + \mathcal{A}/2\,, 0 \leq \varphi < 2\pi\}$ the estimate

$$k_p \leq \frac{\sqrt{2}}{\pi}\left(1 + \frac{\mathcal{A}}{2}\right). \tag{A.4}$$

This bound is only useful for small $\mathcal{A}$ because the gap width is not taken into account. If we combine (A.4) with the bound derived in [42], we obtain

$$k_p \leq \min\left\{\frac{1}{2}\sqrt{1 + \frac{2}{\mathcal{A}}}, \frac{\sqrt{2}}{\pi}\left(1 + \frac{\mathcal{A}}{2}\right)\right\}. \tag{A.5}$$

Theorem A.3 *If $\Omega = \{(r,\varphi)\colon R_i < r < R_o\,, 0 \leq \varphi < 2\pi\}$, then for the Poincaré constant k_p we have the bound*

$$k_p \leq \sqrt{\frac{R_o}{R_i}}\,\frac{R_o - R_i}{\pi}.$$

Proof. Introducing polar coordinates we obtain

$$k_p^2 = \max_{w \in H_0^1(\Omega)} \frac{\int_\Omega |w|^2\, d(x,y)}{\int_\Omega \nabla w \cdot \nabla w\, d(x,y)} = \max_{\tilde{w} \in H_0^1(\Omega)} \frac{\int_\Omega r\,|\tilde{w}|^2\, d(r,\varphi)}{\int_\Omega r\left((\partial_r \tilde{w})^2 + r^{-2}(\partial_\varphi \tilde{w})^2\right) d(r,\varphi)}, \tag{A.6}$$

where $\tilde{w}(r,\varphi) = w(x,y)$. From [71, Proposition 1.1] we know that the argument for which (A.6) attains its maximum value is a radial function $\tilde{u}_1 = \tilde{u}_1(r)$. Hence, there holds

$$k_p^2 = \frac{\int_{R_i}^{R_o} r\,|\tilde{u}_1(r)|^2\, dr}{\int_{R_i}^{R_o} r\,|\tilde{u}_1'(r)|^2\, dr}.$$

Note that $\tilde{u}_1$ is the eigenfunction associated to the smallest eigenvalue of the Laplace problem (A.2). Further, we know from the one-dimensional Laplace eigenvalue problem that

$$\left(\frac{R_o - R_i}{\pi}\right)^2 = \max_{\tilde{w} \in H_0^1(R_i, R_o)} \frac{\int_{R_i}^{R_o} |\tilde{w}(r)|^2\, dr}{\int_{R_i}^{R_o} |\tilde{w}'(r)|^2\, dr},$$

which in combination with (A.1) implies

$$k_p^2 \leq \frac{R_o}{R_i}\,\frac{\int_{R_i}^{R_o} |\tilde{u}_1(r)|^2\, dr}{\int_{R_i}^{R_o} |\tilde{u}_1'(r)|^2\, dr} \leq \frac{R_o}{R_i}\left(\frac{R_o - R_i}{\pi}\right)^2.$$

□

By virtue of Theorem A.3 we already obtain a new bound for the Poincaré constant with

respect to $\Omega_\mathcal{A}$, namely
$$k_p \leq \frac{1}{\pi}\sqrt{1+\frac{2}{\mathcal{A}}}.$$
Together with (A.4) we have
$$k_p \leq \min\left\{\frac{1}{\pi}\sqrt{1+\frac{2}{\mathcal{A}}},\frac{\sqrt{2}}{\pi}\left(1+\frac{\mathcal{A}}{2}\right)\right\}. \tag{A.7}$$

A.2 Evaluation of the Poincaré Constant

In order to perform an efficient evaluation of the Poincaré constant, we want to exploit its one-dimensional character. Therefore, we transform the eigenvalue problem (A.2) to polar coordinates and use the result of [71] that all eigenfunctions associated to the smallest eigenvalue are radial. Hence, we have that $k_p = \lambda_1^{-1/2}$, where λ_1 is the smallest eigenvalue of the one-dimensional problem
$$-\tilde{u}'' - \frac{1}{r}\tilde{u}' = \lambda\tilde{u} \quad \text{in } (R_i, R_o), \qquad \tilde{u}(R_i) = \tilde{u}(R_o) = 0.$$

The subsequent computations were performed in the framework of the finite element software HiFlow, see [1]. First, the problem was discretized by means of a finite difference scheme of second order. Then, the algebraic eigenvalue problem was solved by means of the Davidson method, see Section 5.1. The numerical results as well the bounds (A.5) and (A.7) are plotted in Figure A.1. The performed computations show that the established bound in Theorem A.3 is almost sharp for large $\mathcal{A}$.

A.3 Application in Natural Convection

We consider the steady-state formulation of the Oberbeck-Boussinesq equations (2.8) as in the setup of Section 6.2. By using polar coordinates (r,φ) and the defined dimensionless quantities $\mathcal{A}$, Ra, Pr, we obtain the dimensionless formulation
$$\frac{1}{Pr}(\mathbf{v}\cdot\nabla)\mathbf{v} - \Delta\mathbf{v} + \nabla p = \frac{Ra}{\mathcal{B}}\sin\varphi\,\mathbf{e}_r + Ra\,\tau\,\mathbf{e}_3,$$
$$\nabla\cdot\mathbf{v} = 0, \tag{A.8}$$
$$\mathbf{v}\cdot\nabla\tau - \Delta\tau = \frac{v_r}{r\mathcal{B}}$$

in $\Omega_\mathcal{A}$, with boundary conditions $\mathbf{v} = \mathbf{0}$, $\tau = 0$ on $\partial\Omega_\mathcal{A}$, see [77]. Here, $\tau = \tau(r,\varphi)$ is the excess temperature and we have set $\mathcal{B} = \ln(1+2/\mathcal{A}) = \ln(R_o/R_i)$. Furthermore, $\mathbf{e}_3$ is the unit vector into direction z, i.e. $\mathbf{e}_3 = \sin\varphi\,\mathbf{e}_r + \cos\varphi\,\mathbf{e}_\varphi$.

102 ESTIMATES AND CALCULATIONS FOR THE POINCARÉ CONSTANT

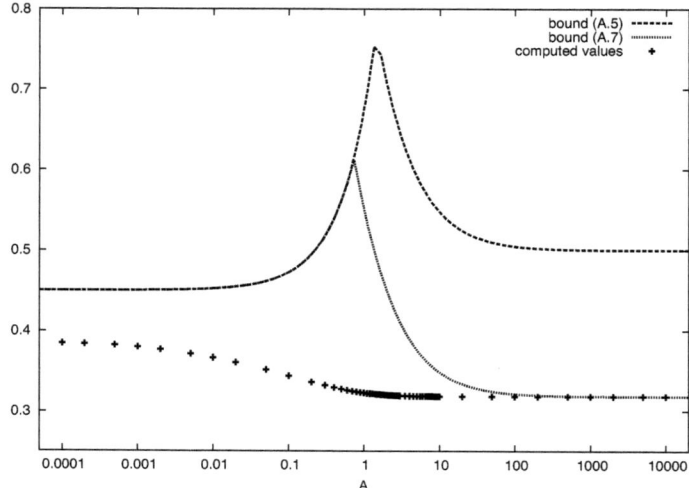

Figure A.1: Evaluation of Poincaré constants with respect to $\mathcal{A}$.

Because no exact solutions are known, simplified models may be more convenient. For instance one may decouple the system by replacing the first equation of (A.8) by a linear equation yielding

$$-\Delta \mathbf{v} + \nabla p = \frac{Ra}{\mathcal{B}} \sin(\varphi) \, \mathbf{e}_r,$$
$$\nabla \cdot \mathbf{v} = 0, \tag{A.9}$$
$$\mathbf{v} \cdot \nabla \tau - \Delta \tau = \frac{v_r}{r\mathcal{B}}$$

in $\Omega_\mathcal{A}$ with zero boundary conditions for $\mathbf{v}$ and τ on $\partial\Omega$. Note that the third equation is still nonlinear. For this system an analytical solution in terms of Bessel functions is known. The model (A.9) is preferred for describing natural convection in case of a small $\mathcal{A}$ (usually $\mathcal{A} < 2.8$).

An approximation scheme to estimate the relative error between the solution of the full system (A.8) and the solution of its simplification (A.9) in terms of a quantity X under consideration is described in [62]. First, an upper bound f_1 on the norm of the absolute error δX is stated:

$$\delta X \leq f_1(k_p, Ra, Pr, \mathcal{A}). \tag{A.10}$$

Second, a lower bound f_2 on the norm of the solution X_0 of the simplified problem is established:

$$f_2(Ra, Pr, \mathcal{A}) \leq X_0. \tag{A.11}$$

A.3. Application in Natural Convection

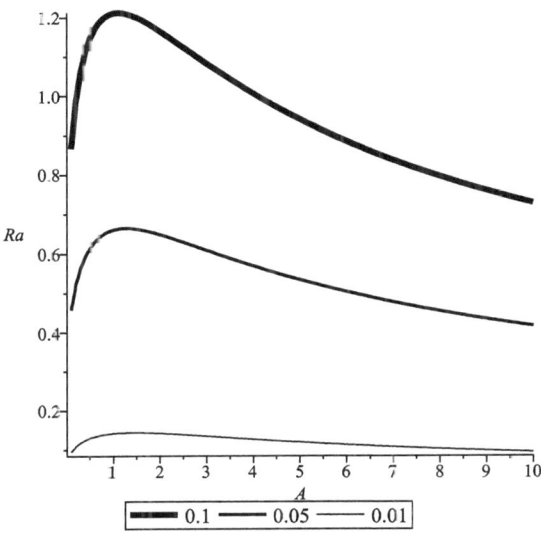

Figure A.2: Contour plots of $\mathcal{E} = 0.01, 0.05, 0.1$.

Finally, by combining (A.10) and (A.11), we obtain

$$\mathcal{E} = \frac{\delta X}{X} \leq \frac{\delta X}{X_0 - \delta X} \leq \frac{f_1}{f_2 - f_1} \quad \text{(A.12)}$$

as an upper bound of the relative error $\mathcal{E}$, provided that $f_2 > f_1$. In terms of $\nabla \mathbf{v}$ and $\nabla \tau$ the functions f_1 and f_2 can be found in [62]. Since these functions are rather technical, we do not state them here. Using the derived estimate (A.7) for the Poincaré constant, we have computed upper bounds for the relative error $\mathcal{E}$ by evaluating f_1 and f_2. In Figure A.2 contour plots of the upper bound (A.12) as a function of Ra and $\mathcal{A}$, with fixed Prandtl number $Pr = 0.7$, in terms of $\nabla \mathbf{v}$ are shown.

Bibliography

[1] HiFlow - a general finite element toolbox. http://www.hiflow.de, 2011.

[2] ALONSO, A., AND DELLO RUSSO, A. Spectral approximation of variationally-posed eigenvalue problems by nonconforming methods. *Journal of Computational and Applied Mathematics 223*, 1 (2009), 177–197.

[3] ALT, H. W. *Lineare Funktionalanalysis*, 5th ed. Springer, 2006.

[4] AMANN, H. *Gewöhnliche Differentialgleichungen*, 2nd ed. de Gruyter, 1995.

[5] AMESTOY, P. R., DUFF, I. S., KOSTER, J., AND L'EXCELLENT, J.-Y. A fully asynchronous multifrontal solver using distributed dynamic scheduling. *SIAM Journal on Matrix Analysis and Applications 23*, 1 (2001), 15–41.

[6] ANZT, H., AUGUSTIN, W., BAUMANN, M., BOCKELMANN, H., GENGENBACH, T., HAHN, T., HEUVELINE, V., KETELAER, E., LUKARSKI, D., OTZEN, A., RITTERBUSCH, S., ROCKER, B., RONNÅS, S., SCHICK, M., SUBRAMANIAN, C., WEISS, J.-P., AND WILHELM, F. HiFlow^3 - a flexible and hardware-aware parallel finite element package. EMCL Preprint Series, 2010.

[7] ARIS, R. *Vectors, tensors, and the basic equations of fluid mechanics*. Dover Publications, 1989.

[8] AULBACH, B. *Gewöhnliche Differentialgleichungen*. Spektrum Akademischer Verlag, 1997.

[9] BABUSKA, I., AND OSBORN, J. Eigenvalue problems. In *Finite Element Methods (Part 1)*, vol. 2 of *Handbook of Numerical Analysis*. Elsevier, 1991, pp. 641–787.

[10] BALAY, S., BUSCHELMAN, K., EIJKHOUT, V., GROPP, W. D., KAUSHIK, D., KNEPLEY, M. G., MCINNES, L. C., SMITH, B. F., AND ZHANG, H. PETSc users manual. Tech. Rep. ANL-95/11 - Revision 3.0.0, Argonne National Laboratory, 2008.

[11] BALAY, S., BUSCHELMAN, K., GROPP, W. D., KAUSHIK, D., KNEPLEY, M. G., MCINNES, L. C., SMITH, B. F., AND ZHANG, H. PETSc web page. http://www.mcs.anl.gov/petsc, 2011.

[12] BALAY, S., GROPP, W. D., MCINNES, L. C., AND SMITH, B. F. Efficient management of parallelism in object oriented numerical software libraries. In *Modern Software Tools in Scientific Computing* (1997), E. Arge, A. M. Bruaset, and H. P. Langtangen, Eds., Birkhäuser Press, pp. 163–202.

[13] BELL, G. I., AND GLASSTONE, S. *Nuclear reactor theory*. Van Nostrand Reinhold, 1970.

[14] BELL, G. I., HANSEN, G. E., AND SANDMEIER, H. A. Multitable treatments of anisotropic scattering in S_N multigroup transport calculations. *Nuclear Science and Engineering, 28* (1967).

[15] BÉNARD, H. Les tourbillons cellulaires dans une nappe liquide. *Revue générale des Sciences pures et appliquées 11* (1900).

[16] BERTSCH, C. Finite-Elemente-Diskretisierung und ein Mehrgitterverfahren für unsymmetrische Eigenwertprobleme. Diplomarbeit, Universität Heidelberg, 1999.

[17] BOLTZMANN, L. Weitere Studien über das Wärmegleichgewicht unter Gasmolekülen. *Wiener Berichte 66* (1872), 275–370.

[18] BOUSSINESQ, J. *Théorie analytique de la chaleur*. Gauthier-Villars, 1903.

[19] BRAESS, D. *Finite Elemente*, 4th ed. Springer, 2007.

[20] BRAMBLE, J. H., AND OSBORN, J. E. Rate of convergence estimates for nonselfadjoint eigenvalue approximations. *Mathematics of Computation 27*, 123 (1973), 525–549.

[21] BRENNER, S. C., AND SCOTT, L. R. *The mathematical theory of finite element methods*, 3rd ed. Springer, 2008.

[22] CAO, Y., AND LEE, J. C. A Krylov subspace method to calculate k- and α- eigenvalues. In *Physics of Reactors "Nuclear Power: A Sustainable Resource" (PHYSOR 2008)* (Interlaken, Switzerland, September 14–19, 2008).

[23] CARPRAUX, J. F., ERHEL, J., AND SADKANE, M. Spectral portrait for non hermitian large sparse matrices. *Computing 53* (1994), 301–310.

[24] CERCIGNANI, C. *The Boltzmann equation and its applications*. Applied mathematical sciences ; 67. Springer, 1988.

[25] CHAITIN-CHATELIN, F., AND HARRABI, A. About definitions of pseudospectra of closed operators in Banach spaces. Tech. Rep. TR/PA/98/08, CERFACS, 1998.

[26] CHANDRASEKHAR, S. *Hydrodynamic and hydromagnetic stability*. Dover Publications, 1981.

[27] CHOI, H., HINZE, M., AND KUNISCH, K. Instantaneous control of backward-facing step flows. *Applied Numerical Mathematics 31*, 2 (1999), 133–158.

[28] CHORIN, A. J., AND MARSDEN, J. E. *A mathematical introduction to fluid mechanics*, 3rd ed. Springer, 2000.

[29] CIARLET, P. G. *The finite element method for elliptic problems*. Classics in applied mathematics ; 40. Society for Industrial and Applied Mathematics, 2002.

[30] CROUZEIX, M., PHILIPPE, B., AND SADKANE, M. The Davidson method. *SIAM Journal on Scientific Computing 15* (1994), 62–76.

[31] DALECKIĬ, J. L., AND KREĬN, M. G. *Stability of solutions of differential equations in Banach space*. Translations of mathematical monographs ; 43. American Mathematical Society, 1974.

[32] DAVIDSON, E. R. The iterative calculation of a few of the lowest eigenvalues and corresponding eigenvectors of large real-symmetric matrices. *Journal of Computational Physics 17* (1975), 87–94.

[33] DEMMEL, J. W., EISENSTAT, S. C., GILBERT, J. R., LI, X. S., AND LIU, J. W. H. A supernodal approach to sparse partial pivoting. *SIAM Journal on Matrix Analysis and Applications 20*, 3 (1999), 720–755.

[34] DESCLOUX, J., NASSIF, N., AND RAPPAZ, J. On spectral approximation. Part 1. The problem of convergence. *RAIRO Analyse Numérique 12*, 2 (1978), 97–112.

[35] DESCLOUX, J., NASSIF, N., AND RAPPAZ, J. On spectral approximation. Part 2. Error estimates for the Galerkin method. *RAIRO Analyse Numérique 12*, 2 (1978), 113–119.

[36] DONGARRA, J., LUMSDAINE, A., NIU, X., POZO, R., AND REMINGTON, K. LAPACK working note 74: A sparse matrix library in C++ for high performance architectures. Tech. rep., 1994.

[37] DRAZIN, P. G. *Introduction to hydrodynamic stability*. Cambridge texts in applied mathematics ; 32. Cambridge University Press, 2002.

[38] DRAZIN, P. G., AND REID, W. H. *Hydrodynamic Stability*, 2nd ed. Cambridge University Press, 2004.

[39] DYKO, M. P., VAFAI, K., AND MOJTABI, A. K. A numerical and experimental investigation of stability of natural convective flows within a horizontal annulus. *Journal of Fluid Mechanics 381* (1999), 27–61.

[40] EMBREE, M., AND TREFETHEN, L. N. Generalizing eigenvalue theorems to pseudospectra theorems. *SIAM Journal on Scientific Computing 23*, 2 (2001), 583–590.

[41] ERN, A., AND GUERMOND, J.-L. *Theory and practice of finite elements*. Applied mathematical sciences ; 159. Springer, 2004.

[42] FERRARIO, C., PASSERINI, A., AND PIVA, S. A Stokes-like system for natural convection in a horizontal annulus. *Nonlinear Analysis: Real World Applications 9*, 2 (2008), 403–411.

[43] FRAYSSE, V., GUEURY, M., NICOUD, F., AND TOUMAZOU, V. Spectral portraits for matrix pencils, 1996.

[44] GALDI, G. P. *An introduction to the mathematical theory of the Navier-Stokes equations*, vol. 1: Linearised steady problems of *Springer tracts in natural philosophy ; 38*. Springer, 1998.

[45] GILL, D. F., AND AZMY, Y. Y. Jacobian-free Newton-Krylov as an alternative to power iterations for the k-eigenvalue transport problem. *ANS Transactions 92* (2005), 728–730.

[46] GILL, D. F., AND AZMY, Y. Y. A Jacobian-free Newton-Krylov iterative scheme for criticality calculations based on the neutron diffusion equation. In *Mathematics, Computational Methods & Reactor Physics (M&C 2009)* (Saratoga Springs, New York, USA, May 3–7, 2009).

[47] GIRAULT, V., AND RAVIART, P.-A. *Finite element methods for Navier-Stokes equations*. Springer series in computational mathematics ; 5. Springer, 1986.

[48] GODUNOV, S. K., AND SADKANE, M. Computation of pseudospectra via spectral projectors. *Linear Algebra and its Applications 279*, 1–3 (1998), 163–175.

[49] GOLUB, G. H., AND VAN LOAN, C. F. *Matrix computations*, 3rd ed. Johns Hopkins University Press, 1997.

[50] HACKBUSCH, W. *Elliptic differential equations*. Springer series in computational mathematics ; 18. Springer, 2003.

[51] HAHN, W. *Stability of motion*. Die Grundlehren der mathematischen Wissenschaften in Einzeldarstellungen ; 138. Springer, 1967.

[52] HAVET, M. *Solution of Algebraic Problems arising in Nuclear Reactor Simulations using Jacobi-Davidson and Multigrid Methods*. PhD thesis, Université Libre de Bruxelles, Belgium, 2008.

[53] HEUVELINE, V. Finite element approximations of eigenvalue problems for elliptic partial differential operators. Habilitationsschrift, Universität Heidelberg, 2002.

[54] HEUVELINE, V., PASSERINI, A., SUBRAMANIAN, C., AND THÄTER, G. Estimates and calculations for the Poincaré constant in an annulus. EMCL Preprint Series, to appear.

[55] HEUVELINE, V., PHILIPPE, B., AND SADKANE, M. Parallel computation of spectral portrait of large matrices by Davidson type methods. *Numerical Algorithms 16*, 1 (1997), 55–75.

[56] HEUVELINE, V., AND RANNACHER, R. A posteriori error control for finite element approximations of elliptic eigenvalue problems. *Advances in Computational Mathematics 15*, 1 (2001), 107–138.

[57] HEUVELINE, V., AND RANNACHER, R. Adaptive FEM for eigenvalue problems with application in hydrodynamic stability analysis. *Journal of Numerical Mathematics* (2006), 1–32.

[58] HEUVELINE, V., SUBRAMANIAN, C., LUKARSKI, D., AND WEISS, J.-P. A multi-platform linear algebra toolbox for finite element solvers on heterogeneous clusters. In *IEEE Cluster 2010, Workshop on Parallel Programming and Applications on Accelerator Clusters (PPAAC 2010)* (Heraklion, Greece, September 20–24, 2010).

[59] KATO, T. *Perturbation theory for linear operators*. Die Grundlehren der mathematischen Wissenschaften in Einzeldarstellungen ; 132. Springer, 1966.

[60] KNOLL, D. A., PARK, H., AND SMITH, K. Application of the Jacobian-free Newton-Krylov method in computational reactor physics. In *Mathematics, Computational Methods & Reactor Physics (M&C 2009)* (Saratoga Springs, New York, USA, May 3–7, 2009).

[61] KOLATA, W. G. Approximation in variationally posed eigenvalue problems. *Numerische Mathematik 29* (1978), 159–171.

[62] LAMACZ, A., PASSERINI, A., AND THÄTER, G. Natural convection in horizontal annuli: evaluation of the error for two approximations. Submitted for publication.

[63] LATHOUWERS, D. Computing time-eigenvalues using the even-parity transport form. *Annals of Nuclear Energy 33*, 10 (2006), 941–943.

[64] LAVALLÉE, P.-F., AND SADKANE, M. Computation of pseudospectra by spectral dichotomy methods in a parallel environment. *Numerical Algorithms 33*, 1 (2003), 343–355.

[65] LEWIS, E. E. Second-order neutron transport methods. In *Nuclear Computational Science*, Y. Azmy and E. Sartori, Eds. Springer, 2010, pp. 85–115.

[66] LEWIS, E. E., AND MILLER, W. F. *Computational methods of neutron transport*. Wiley, 1984.

[67] LI, X. S., AND SHAO, M. A supernodal approach to incomplete LU factorization with partial pivoting. *ACM Transactions on Mathematical Software 37*, 4 (2010).

[68] MERCIER, B., OSBORN, J., RAPPAZ, J., AND RAVIART, P. A. Eigenvalue approximation by mixed and hybrid methods. *Mathematics of Computation 36*, 154, 427–453.

[69] MORGAN, R. B. Generalizations of Davidson's method for computing eigenvalues of large nonsymmetric matrices. *Journal of Computational Physics 101*, 2 (1992), 287–291.

[70] MORGAN, R. B., AND SCOTT, D. S. Generalizations of Davidson's method for computing eigenvalues of sparse symmetric matrices. *SIAM Journal on Scientific and Statistical Computing 7*, 3 (1986), 817–825.

[71] NAZAROV, A. I. The one-dimensional character of an extremum point of the Friedrichs inequality in spherical and plane. *Journal of Mathematical Sciences 102*, 5 (2000), 4473–4486.

[72] NOTAY, Y. Is Jacobi-Davidson faster than Davidson? *SIAM Journal on Matrix Analysis and Applications 26*, 2 (2005), 522–543.

[73] OBERBECK, A. Über die Wärmeleitung der Flüssigkeiten bei Berücksichtigung der Strömungen infolge von Temperaturdifferenzen. *Annalen der Physik 243* (1879), 271–292.

[74] OECD/NEA. Benchmark on deterministic transport calculations without spatial homogenisation. A 2-D/3-D MOX fuel assembly benchmark. In *NEA/NSC/DOC(2003)16* (2005), Nuclear Science, OECD, Nuclear Energy Agency.

[75] OSBORN, J. E. Spectral approximation for compact operators. *Mathematics of Computation 29*, 131 (1975), 712–725.

[76] OSBORN, J. E. Approximation of the eigenvalues of a nonselfadjoint operator arising in the study of the stability of stationary solutions of the Navier-Stokes equations. *SIAM Journal on Numerical Analysis 13*, 2 (1976), 185–197.

[77] PASSERINI, A., FERRARIO, C., RŮŽIČKA, M., AND THÄTER, G. Theoretical results on steady convective flows between horizontal coaxial cylinders. *SIAM Journal on Applied Mathematics* (submitted).

[78] PASSERINI, A., RŮŽIČKA, M., AND THÄTER, G. Natural convection between two horizontal coaxial cylinders. *Zeitschrift für Angewandte Mathematik und Mechanik 89*, 5 (2009), 399–413.

[79] PAZY, A. *Semigroups of linear operators and applications to partial differential equations*, 2nd ed. Applied mathematical sciences ; 44. Springer, 1992.

[80] PHILIPPE, B., AND SADKANE, M. Computation of the fundamental singular subspace of a large matrix. *Linear Algebra and its Applications 257* (1997), 77–104.

[81] QUARTERONI, A., AND VALLI, A. *Numerical approximation of partial differential equations*. Springer series in computational mathematics ; 23. Springer, 1994.

[82] RANNACHER, R. Numerische Mathematik 3 - Numerik von Problemen der Kontinuumsmechanik. Vorlesungsskriptum, 2006.

[83] RANNACHER, R., AND TUREK, S. A simple nonconforming quadrilateral Stokes element. *Numerical Methods for Partial Differential Equations 8* (1992), 97–111.

[84] RIEDEL, K. S. Generalized epsilon-pseudospectra. *SIAM Journal on Numerical Analysis 31*, 4 (1994), 1219–1225.

[85] RUHE, A. The rational Krylov algorithm for large nonsymmetric eigenvalues – mapping the resolvent norms (pseudospectrum), 1994.

[86] SAAD, Y. *Numerical methods for large eigenvalue problems*. Manchester University Press, 1992.

[87] SAAD, Y. *Iterative methods for sparse linear systems*, 2nd ed. Society for Industrial and Applied Mathematics, 2003.

[88] SADKANE, M. Block-Arnoldi and Davidson methods for unsymmetric large eigenvalue problems. *Journal of Numerical Mathematics 64* (1993), 195–211.

[89] SATTINGER, D. The mathematical problem of hydrodynamic stability. *Journal of Mathematics and Mechanics 19*, 9 (1970), 797–817.

[90] SCHMID, P. J., AND HENNINGSON, D. S. *Stability and transition in shear flows*. Applied mathematical sciences 142. Springer, 2001.

[91] SCOTT, R. Interpolated boundary conditions in the finite element method. *SIAM Journal on Numerical Analysis 12*, 3 (1975), 404–427.

[92] SLEIJPEN, G. L. G., AND VAN DER VORST, H. A. A Jacobi-Davidson iteration method for linear eigenvalue problems. *SIAM Review 42*, 2 (2000), 267–293.

[93] SOBOLEV, S. L. *Applications of functional analysis in mathematical physics*. Translations of mathematical monographs ; 7. American Mathematical Society, 1963.

[94] STRAUGHAN, B. *The energy method, stability, and nonlinear convection*. Applied mathematical sciences ; 91. Springer, 1992.

[95] STRAUSS, W. A. *Partial differential equations*. Wiley, 1992.

[96] SUBRAMANIAN, C., VAN CRIEKINGEN, S., HEUVELINE, V., NATAF, F., AND HAVÉ, P. The Davidson method as an alternative to power iterations for criticality calculations. Submitted for publication.

[97] SUBRAMANIAN, C., VAN CRIEKINGEN, S., HEUVELINE, V., NATAF, F., AND HAVÉ, P. The Davidson method as an alternative to power iterations for criticality calculations. In *Mathematics and Computational Methods applied to Nuclear Science and Engineering (MC 2011)* (Rio de Janeiro, Brazil, May 8–12, 2011, submitted).

[98] SÜLI, E., AND MAYERS, D. F. *An introduction to numerical analysis*. Cambridge University Press, 2003.

[99] TAKEDA, T., AND IKEDA, H. Final report on 3D neutron transport benchmark. Department of Nuclear Engineering, Osaka University, Japan, 1991.

[100] TAKEDA, T., TAMITANI, M., AND UNESAKI, H. Proposal of 3D neutron transport benchmark. *NEACRP A-953 REV. 1* (1989).

[101] TOH, K.-C., AND TREFETHEN, L. N. Calculation of pseudospectra by the Arnoldi iteration. *SIAM Journal on Scientific Computing 17*, 1 (1996), 1–15.

[102] TREFETHEN, L. N. Computation of pseudospectra. *Acta Numerica 8* (1999), 247–295.

[103] TREFETHEN, L. N., AND EMBREE, M. *Spectra and pseudospectra*. Princeton University Press, 2005.

[104] TREFETHEN, L. N., TREFETHEN, A. E., REDDY, S. C., AND DRISCOLL, T. A. Hydrodynamic stability without eigenvalues. *Science 261*, 5121 (1993), 578–584.

[105] TRITTON, D. J. *Physical fluid dynamics*. The modern university physics series. Van Nostrand Reinhold, 1979.

[106] VAN CRIEKINGEN, S. A 2D/3D cartesian geometry non-conforming spherical harmonic neutron transport solver. *Annals of Nuclear Energy 34*, 3 (2007), 177–187.

[107] VAN CRIEKINGEN, S., NATAF, F., AND HAVÉ, P. Parafish: a parallel $FE - P_N$ neutron transport solver based on domain decomposition. *Annals of Nuclear Energy 38*, 1 (2011), 145–150.

[108] VAN DORSSELAER, J. L. M. Pseudospectra for matrix pencils and stability of equilibria. *BIT Numerical Mathematics 37* (1997), 833–845.

[109] VAN DORSSELAER, J. L. M. Several concepts to investigate strongly nonnormal eigenvalue problems. *SIAM Journal on Scientific Computing 24*, 3 (2002), 1031–1053.

[110] VERDÚ, G., GINESTAR, D., MIRÓ, R., AND VIDAL, V. Using the Jacobi-Davidson method to obtain the dominant lambda modes of a nuclear power reactor. *Annals of Nuclear Energy 32* (2005), 1274–1296.

[111] VERDÚ, G., MIRÓ, R., GINESTAR, D., AND VIDAL, V. The implicit restarted Arnoldi method, an efficient alternative to solve the neutron diffusion equation. *Annals of Nuclear Energy 26* (1999), 579–593.

[112] WARSA, J. S., WAREING, T. A., MOREL, J. E., MCGHEE, J. M., AND LEHOUCQ, R. B. Krylov subspace iterations for deterministic k-eigenvalue calculations. *Nuclear Science and Engineering 147*, 1 (2004), 26–42.

[113] WERNER, D. *Funktionalanalysis*, 6th ed. Springer, 2007.

[114] WESSELING, P. *Principles of computational fluid dynamics*. Springer series in computational mathematics ; 29. Springer, 2009.

[115] WILLIAMS, W. S. C. *Nuclear and particle physics*. Oxford science publications. Clarendon Press, 1992.

[116] WLOKA, J. *Partial differential equations*. Cambridge University Press, 1992.

[117] YOO, J.-S. Dual steady solutions in natural convection between horizontal concentric cylinders. *International Journal of Heat and Fluid Flow 17*, 6 (1996), 587–593.

[118] YOO, J.-S. Natural convection in a narrow horizontal cylindrical annulus: $Pr \leq 0.3$. *International Journal of Heat and Mass Transfer 41*, 20 (1998), 3055–3073.

[119] YOO, J.-S. Transition and multiplicity of flows in natural convection in a narrow horizontal cylindrical annulus: $Pr = 0.4$. *International Journal of Heat and Mass Transfer 42*, 4 (1999), 709–722.

[120] YOSIDA, K. *Functional analysis*, 4th ed. Die Grundlehren der mathematischen Wissenschaften in Einzeldarstellungen ; 123. Springer, 1974.

I want morebooks!

Buy your books fast and straightforward online - at one of world's fastest growing online book stores! Environmentally sound due to Print-on-Demand technologies.

Buy your books online at
www.morebooks.shop

Kaufen Sie Ihre Bücher schnell und unkompliziert online – auf einer der am schnellsten wachsenden Buchhandelsplattformen weltweit! Dank Print-On-Demand umwelt- und ressourcenschonend produziert.

Bücher schneller online kaufen
www.morebooks.shop

KS OmniScriptum Publishing
Brivibas gatve 197
LV-1039 Riga, Latvia
Telefax +371 686 20455

info@omniscriptum.com
www.omniscriptum.com

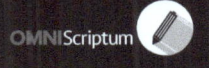

Printed by Books on Demand GmbH, Norderstedt / Germany